DARKENING OF THE LIGHT

Witnessing the End of an Era

LLEWELLYN VAUGHAN-LEE

THE GOLDEN SUFI CE[

DARKENING
of the LIGHT
Witnessing the End of an Era

LLEWELLYN VAUGHAN-LEE

First published in the United States in 2013 by
The Golden Sufi Center
P.O. Box 456, Point Reyes, California 94956
www.goldensufi.org

ISBN: 978-1-890350-50-5

Printed and Bound by Thomson-Shore, Inc.

Library of Congress Cataloging-in-Publication Data

Vaughan-Lee, Llewellyn.
  Darkening of the light : witnessing the end of an era / Llewellyn
Vaughan-Lee.
    pages cm
  Includes bibliographical references.
  ISBN 978-1-890350-50-5 (pbk. : alk. paper)
1. Sufism. I. Title.
  BP189.2.V38 2013
  297.4--dc23
                              2013026010

CONTENTS

Introduction   i

1. A New Light (2004)   11

2. The Awakening of the World (2006)   27

3. A Return to What Is Simple (2006)   37

4. A Prayer for the World (2008)   45

5. Waiting for the Dawn (2009)   63

6. The Darkness Before the Dawn (2009)   69

7. What Happens When the Ice Melts (2010)   79

8. The Wall (2010)   89

9. Witnessing the End of an Era (2011)   111

10. Hungry Ghosts and the   129
    Sacred Substance in Creation (2011)

11. Darkness and Light (Summer 2012)   135

12. A Story of a Lost Opportunity (Fall 2012)   143

Epilogue: A Winter Solstice (Winter 2012)   151

Notes   158
Acknowledgments   164

The light has sunk into the earth:
The image of Darkening of the Light.

I CHING – HEXAGRAM 36

INTRODUCTION

There is a spiritual tradition of watching and witnessing: that events in the inner and outer worlds need to be witnessed. This book is a witness to changes that have been happening in the inner worlds[1] over the last few years, changes that are directly reflected by those in the outer world that belong to our growing sense of global interconnectedness and an awakening consciousness of oneness. In the outer world we can witness the effects of this unifying consciousness in the dawning recognition that we are one global human community and also part of one living interdependent ecosystem. The shadow-side of this awareness is an expanded global exploitation together with an accelerating ecological imbalance that some say has now pushed our planet's ecosystem past its "tipping point," with unforeseeable consequences.

Spiritual teachings tell us that events in the outer world are a reflection of changes taking place in the inner worlds. There have been changes in the inner world as vital to our collective destiny as what has been happening on the outer global stage, and yet these inner changes have been mostly unreported, unacknowledged. From within the inner world there has come an energy shift that belongs to this global awakening and evolution in consciousness: humanity's transition into a consciousness of oneness. But there has also been a darkening of the light that was a catalyst for these changes—a darkening that directly reflects our continuing ecological devastation.

Over the last decade or more we have heard about the need and possibility for a "paradigm shift," for a global shift in consciousness: how humanity is at the dawn of a new age. I first became fully aware of this possibility at the beginning of the new millennium, when in the spring of 2000, I noticed a new energy present. This energy had a spark, a quality of joy, and a deep knowing of oneness. It belonged to the whole of life and to the Earth itself, as well as to the heart and soul of humanity. It had within it all of the seeds of the next era of human evolution—qualities of oneness, interconnectedness, and the simple joy of life. This was a spark to help the world to awaken, to step out of an era of separation into an era of divine presence and unity.

This spark carried the possibility that the world could come alive in a new way, awaken to its magical potential, and that the whole of humanity could be part of this awakening. In simplicity and a certain naïveté I saw that there was a oneness waiting to be lived, and that we were this oneness. And yet at the same time, in order for this opportunity to be realized, humanity had to leave behind certain behavior patterns and power structures that resisted this new energy, that wanted the old ways to remain.

This new energy came into the world in the form of a spark of light to help humanity to evolve, to help it to grow out of an era dominated by a sense of separation, particularly separation from the world around us. It was a catalyst to enable the last era to transform, to help a new global awareness to be born. And this light was given freely: nothing was asked in return. And yet, as with all gifts, there was an unspoken responsibility: that this light be used for its true purpose.

But this gift of light was also a test for humanity: to recognize the oneness of creation and its sacred nature, to turn away from the darkness of a civilization that celebrates the ego and its desires, and has forgotten our ancient purpose as guardians of the planet. It might appear contradictory that the journey towards oneness means confronting the duality of light and darkness, and choosing the light over the darkness, but this is the reality we are experiencing. On our present global stage the forces of darkness in the forms of unrestrained materialism, greed, and other expressions of self-interest are destroying our ecosystem, and in order to embrace the consciousness of oneness that values the wonder and diversity of creation, we need to become free from their grip and from the ego-centered values that encourage them.[2] We have to make the simple but vital shift from "me" to "we."

And slowly, gradually, for a time it seemed this shift was taking place: for a time the darkness dissolved and there was hope in the planet; there was a spark of remembrance, an awakening of an ancient joy, the joy that belongs to life when it is recognized as sacred, before the time of division, when there is the knowing that all is One. And gradually a spring came again, and the trees and the animals laughed and thought for a time that their voices would be heard. And with this spring, the knowing returned, the ancient knowing that includes all of life. And all this was welcomed.

And on the human stage there was the beginning of a spiritual reawakening. Over the last decades in the West we have been given greater access to spiritual techniques and practices to help us step out of the world of the ego into the bigger, unified dimension of the Self. Spiritual knowledge that had been hidden for centuries began to be returned to us: the knowledge of shamans and yogis, the wisdom of mystics and monks. Spiritual disciplines that had been practiced in secret became accessible, books that once were hidden in remote libraries were readily available. We began once again to have ways to access the spiritual knowledge of our own soul. The idea of "spiritual awakening" no longer belonged to just a select few, but became a possibility for many people, and even for the world itself. In some countries spiritual persecution remained, but in the West there was a spiritual renaissance that had not been seen for many centuries. The signs of spiritual transformation were more and more visible like buds in springtime. And the spark of light speeded up this transformation, helped life's magic to return, its real spiritual nature to become more visible.

But the darkness of human desire fought back, together with the power structures that did not want any change. The darkness of possessiveness and the evil of greed now walked with heavy boots over the face of the whole world. Globalization meant more exploitation rather than unity. Even the bright lights of spiritual awakening were obscured by the buying and selling of the marketplace—what should be given freely was commercialized and exploited. And slowly the light started to go out. And no one seemed to notice. Everyone was so busy, their attention was lost, caught in the many distractions and demands of daily life. No one seemed to notice that the joy was retreating, that the hope was disappearing. The trees noticed, the birds noticed, the rivers knew of the light being lost. But for so many centuries we had forgotten how to listen to the trees, how to hear the birds or the rivers. We no longer knew how to read the signs. And so we did not notice what was happening: how the light that was given was being swallowed up, how a spark was becoming lost. And many voices tried to speak to humanity, voices within nature, voices of indigenous peoples and others. But these voices were not heard, or if they were heard they were ignored. And so the light receded until its spark was finally lost. And without this spark, this light, there could be no real change: nothing new could be born. And this is the great unspoken tragedy of today. This is what has happened to the soul of humanity and the soul of the world.

But people behave as if nothing has happened. This for me is a great mystery, this complete denial of what was given and what was lost. People continue with their daily affairs, their buying and selling, their dreams and desires. And they seem to notice nothing, and yet in their souls something is missing. A note is no longer present, a color in the spectrum of the soul is missing, a joy has fled. And now the future is very different—because that note, that color, would have allowed something to happen that cannot now take place. Instead humanity must continue without a certain dream being born, without a certain wonder being made present, without a quality of life's oneness coming alive. Humanity must continue in this way because it did not take responsibility for what was given—it did not remember.

In the legends of the Hopi people, in three previous worlds humanity forgot the plan of the Creator. Three previous worlds were destroyed: the first with fire, the second with ice, and the third with floods. In some stories a few people were led from the old, dying world to the next world.

And now, as so many species die and the world becomes more and more toxic, it seems we are coming to the end of the fourth world.

People are now acknowledging that we are at the end of an era. The Cenozoic Era, which has lasted for the last sixty-five million years, is coming to an end with the present mass depletion of species, a depletion which for the first time in our world's history is man-made. But there is another tragedy that is not being told: how humanity has once again missed an opportunity; how a possibility for a collective shift in consciousness and a global transformation has been lost.

In many, many ways, this is a story I would rather not tell. It also has a very personal note, in that I have been deeply drawn into this unfolding drama since I first felt the new energy arriving. Then there was a primal hope and joy, the sense of something within the world that was finally awakening, and a deep knowing of the possibilities it could bring. I wrote a series of six books on how to welcome and work with this energy, how it could change our consciousness and our world.[3] It became central to my inner life. Then, as I watched the possibilities for a real global awakening of oneness fade away in this darkening light, it brought a terrible sadness to my soul.

I would have rather remained in innocence or ignorance. It is much easier to say that everything is all right, that humanity is in the midst of a paradigm shift, that we are all awakening to oneness, that the darkness around us is just part of the process. And of course in some ways this is true. There is an awakening, and there is a darkness that is resistance. There are the old ways that are dying, just as something new becomes visible. Particularly in the West many individuals have awakened to their spiritual nature. We can recognize the unity and interconnectedness of life more easily than even a decade ago—it is even mirrored in the Internet, whose instantaneous patterns of connection have suddenly become part of the fabric of our daily life in so many ways.[4] We are more and more part of an interrelated human network.

We are one world: financially, socially, and ecologically interdependent. But there is no joy present in these interconnections, no sense that life itself is coming alive in a new way. The possibility for a real global shift has passed: humanity has failed to take responsibility for creation. Instead of real unity we are creating more human inequality, more exploitation, and an ever greater ecological imbalance. Once again we have forgotten our role as guardians of the planet. And so something has died—silently, unnoticed. A spark that was present at the dawning of the millennium is now absent. And without this light our present civilization cannot transform.

This small book is a witness, telling the story of this silent tragedy in a series of articles written over the last few years, from when the hope was still present, to the moment in which something died within the soul of the world. The epilogue tells of the winter solstice in 2012, when the 26,000-year cycle prophesied by the Mayans ends, suggesting the possibility of another story being born. The articles are chronologically presented, each chapter like a snapshot on this journey, a different vista of the inner and outer world. Now that this opportunity has come and gone, maybe, as in the story of the Hopi, the time has come for some of us to begin the next journey to the fifth world, when we leave behind the world we have destroyed with our greed and desire. The end of one cycle may point to this new beginning. But however the future may unfold, there is a need to recognize what has happened: to tell the story of a missed opportunity. We need to grieve for what we have lost.

At the beginning of the millennium I became aware of an awakening in the core of the world, a new light that carried the seeds of a possible global transformation. This light and its potential for change needed our attention and participation.

A NEW LIGHT
(2004)

May we be those who shall renew this existence.

ZARATHUSTRA

At the core of creation something is changing, coming alive in a new way. A light at the center of the world that has for millennia been dormant has been rekindled. This is the light of life itself waking up, remembering its own real nature and divine purpose. And with this awakening, the living being that is our world is undergoing a transformation in its very essence. The awakening of the light at the center of the world carries the potential for a whole new revelation, the possibility of a new way of living and being and relating to one another and to life. At this moment, we stand on the edge of a new stage in the evolution of life and consciousness, a new paradigm for the world.

Life is one, has always been one. It is a single, living, organic wholeness and everything in creation is a part of it, as vitally and inseparably related to the whole of life as an individual cell or organ is related to the larger organism of which it is a part. Life is alive, with its own intelligence and divine purpose. And our role, as an intelligent part of this living oneness, is to support and help life to celebrate and fulfill this purpose. We are the guardians of life's sacred meaning.

But we have forgotten this. Our present world bears witness not to the Divine but to our own hubris, our greed, our delusions of power. In our obsessive desire for material comforts, for more and more *things*, we have plundered the Earth as if it were a mere commodity to be exploited, and we have desecrated our world. We live as if life were something apart from ourselves that we can master and control, rather than part of our very being. We have lost all sight of life's real nature and purpose, and our relationship to this purpose. And life has allowed this to happen. Until now life has for the most part remained receptive—waiting, watching us follow our desires, and saying very little, even as it has felt the sorrow of a people who deny its divine nature. But now something deep within life is changing; an era is ending, and with it life's passivity in the face of the abuse we have visited upon it. Life is waking up. And the light of this awakening is extremely dynamic. It is metamorphosing very quickly, bringing into its orbit other energies, other forces within creation, some awake, some still dormant, and is setting energies in motion that are going to change our world.

Few have recognized the signs of this change. Most are too busy, too caught up in their own affairs to notice something so new. And those who are fixed on the image of an Earth without a soul will not recognize this new light, will not allow it into their lives. It would be too threatening. Yet the awakening light is beginning to affect human consciousness. It is sending messages of hope into our collective soullessness. And it is attracting those who want to work with it.

This light needs human consciousness to work with it so that it can become fully alive and realize its potential. It needs us to be awake at this moment of awakening, to be alive at this moment of rebirth. And it calls to each of us in our own way, reminding us of an ancient pledge to witness and then participate in what is being born. It is our remembrance and our awakening. It is the light within us. And yet it is also something other, something new that is being given to this world.

Stepping into the arena of this light we become a part of what is being born: we contribute our individual consciousness to this work, our light to this awakening. And we also hold back the forces of darkness, the demands of greed, fear, and power that would otherwise try to abort this new birth, to try to ensure that nothing can be born. In our prayers and our presence we protect this new birth, this awakening of the light of the world. This is a simple commitment from our soul to the soul of the world, an acknowledgment of its need for us and our need for what will be born.

At present we do not easily recognize this light—it is too new to be stored in our memories. It has no image in our minds. And yet it speaks to us, continually trying to draw our attention. For now it still speaks to us in its own language, in the ways of light that are almost invisible to our human consciousness. It has not yet learned the language of duality and separation—it is like a newborn child whose consciousness has not learned the words or divisions of our world. It does not know of our conflicts, even though it can see the scars in our souls. It can sense the obscuration of our own light and the aura of our forgetfulness of the sacred. But it is still too new to know how to communicate in either words or images. At this initial stage it needs us to be present in its inner world of light—a light where there is no real knowing of our darkness.

It needs us to step out of our world of duality and meet it with the light of our own spiritual light. To meet the light in its realm of light means that we must step outside of so much of what we consider important—and not only our own dramas, but also the conflicts and suffering of our world. It means we must be attentive to what can be born rather than obsessed with a self and a civilization that are dying. And yet we must not replace the illusions of this world with a spiritual illusion, a fantasy of escaping the difficulties we face in life. To do that is also to deny the light. For this light that is calling to us from its inner realm of light belongs to life itself. It is a light born from the depths of being, from the core of what is. It is the yes of creation.

To step into its realm we need to leave behind the dramas and conflicts that belong to our ego's version of our self and the world. We need to meet it with our own light, that pure awareness within us that belongs to a higher dimension of our being. In this meeting we will recognize that it is the same light. We are not separate from life, and the light of the world lives within us. We *are* the light of the world; we carry the consciousness of life's highest purpose within the center of our true being. It is present in our every breath. It is our remembrance of the Divine, our essential nature breathing the primal joy of this world. The awakening of the light of the world is our own awakening, as we wake up to our own essential nature and purpose, our belonging to life and to our Creator, which we have for so long forgotten.

CO-CREATING A NEW WORLD

But it is also something completely new, a new energy and consciousness that are being given to this world at this moment in its evolution. And when we bring the light of our consciousness to the light of what is coming into being, we become a partner in its creation. This is real co-creation: the interaction of light with light, our individual light and the world's light, the dynamic interplay of the one light of divine love.

Our true light comes from the divine spark we carry within us. This is the divine spark with which we are born, a spark that carries the deepest purpose of our life. Every child carries this light; it is visible in a child's eyes, in its laughter and joy. But as we grow older, it "fades into the common light of day."[1] Then we must reclaim and reignite its spark through the sometimes painful path of spiritual reawakening. This is our real human consciousness which belongs to the consciousness of our divine nature. And it is far more powerful than we realize. We might have glimpsed some of the ways our consciousness shapes our own reality, perhaps observed the way a simple change in attitude can alter our circumstances. But we have not yet understood that it is at once a catalytic and a determining force in all of creation.

Now we are being asked to wake up to our spiritual self and to assume our real co-creative role in the evolution of the world. With the awakening of the world's light, the inner structure of life itself is changing, releasing primal energies of creation that have not been active for millennia. As they come into play again, they give rise to a whole new range of possibilities, new patterns through which life can flow and the Divine can reveal Itself in the world. These are the patterns that will determine the fundamental structures of the next era, the shape of our future. These primal energies are still in flux; they have not yet constellated into the new patterns. By bringing our consciousness—that awareness in us that is attuned to life's and our own highest purpose—

to these primal energies of creation, we can influence the patterns they will ultimately form; we can help determine the structures that will be born from them, the archetypal images that will shape life in the coming era. We are needed to help create the paradigm within which the awakening world will come alive.

THE POWER OF RECOGNITION

How do we do this work? The first step is a simple recognition of the real nature of life. We need to let go of our old images of life as something defined and apart from ourselves and recognize that life itself is alive, is a dynamic organic being with its own consciousness and intelligence, and that we are a part of it. Simple as it is, this recognition is extremely powerful. It is only when we recognize that life is alive that we can access the light at the core of creation, because that light is the light of life waking up to its own aliveness. And it is only through the dialogue of our light with its light that we can gain access to the primal energies of creation. If we persist in our old images of life, we will never have access to the sacred dimension where the real co-creative work can take place.

From within the world there is a cry for us to help it in this work. The world needs us to recognize it as a living being; it needs the light of our consciousness to interact with the light awakening in its core. Through our light we can

help free the world of the negative patterns that are draining its life energy, and help this energy to flow into the new
patterns that belong to the future. This is a simple act of love
and service through which we acknowledge the real wonder
of the world into which we have been born, the mystery of
incarnation in which we each bring a spark of divine light
into the world where it can help the world to evolve.

Once we acknowledge and then recognize the world as a
living being with its own light, then we can be present where
life comes into being, and it is here that we can participate,
where our spiritual awareness can be present. Without this
conscious awareness of creation as a living being, we stand
on the periphery, an unconscious spectator of our own destiny and the destiny of the world. Taking responsibility for
our planet does not mean just an ecological awareness, but
a recognition that the light awakening in the world needs
our presence and participation.

Every moment the light of the world is changing, evolving. It is the dynamic core of life that carries within it the
secret of creation, of the word of the Divine made manifest.
Just as our light carries the real meaning of our life, the
purpose of our soul, and the knowing of how to live this
purpose, so does the light within the world know the world's
real meaning, its soul's purpose, and what is needed to live
it. This light is ancient and yet new, a continual response to
the divine moment. It carries the real history of the world
and the real knowing of the purpose of this wonder in which

we live. At this time in the world's evolution this wonder is becoming more conscious, and it needs our light to bring its hidden meaning into the consciousness of the world and of humanity.

IMAGES OF LIFE

Through the simple recognition of life's real nature as we go about our ordinary lives, we open ourselves and the world around us to dimensions within life that have been veiled from us by all the layers of projections, our patterns of denial. We have all but forgotten that life is a rich and mysterious coming together of many worlds. We have lost sight of what the ancient priestesses and shamans knew, that the forms of our visible world have their roots in unseen dimensions, and that it is in these unseen dimensions that the primal energies of life lie. In our forgetting, we have lost the wholeness of life, and we have cut ourselves off from the real forces that shape our world. But when we are present in life, free from demands or agendas, when we allow life to unfold according to its own inner principles, we open up a doorway again between the worlds. Within our consciousness the inner and outer, the visible and the unseen worlds, can come together and speak to each other, and our split-apart world can become whole again.

The reconnecting of the worlds is central to the re-awakening of life. For it is in the unmanifest symbolic realm that the primal energies of life constellate into the patterns that shape our lives. When we recognize the meeting of the worlds within us, we regain access to this realm and the powers at work there. We ourselves become the place where the worlds meet, where the primal powers of life, the hidden archetypes, emerge into form. This is the place where the patterns and images that shape the collective life of an era come into being. Standing at that meeting point in ourselves, we can use our awareness to influence their shape and direction.

At this time of momentous change, our consciousness is needed to help midwife these images. This is a time of enormous potential, but also of frightening uncertainty, as the old patterns of life dissolve and the new ones have yet to form, and potent new energies are released. The light at the core of creation is *life coming alive*. It has the violence and power that belong to life. Despite all our illusions of control, we cannot control or contain it—nor would we want to: its primal potency is needed to revitalize the world. We are not here to control life; we are here to witness and revere the Divine as It manifests in a new way through the natural force of life. We need to bring the light of our consciousness to work together with the awakening light within life, so that these primal energies can be creatively channeled, and they do not unleash too much destruction and chaos. Through our consciousness, we can help direct these energies so that they flow into images of life that support the whole of life, both the visible and the unseen worlds, physical life and the life of the soul—so that they reflect life's oneness, its real nature and divine purpose.

Part of this work is simply to be awake, alert to the images emerging both inside us and in the world around us. Some have already appeared, but unless we recognize them for what they are and welcome them consciously into our world, they will not be able to achieve their full potential. These emerging images are much more than pictures of our world to help us understand and navigate it. They come from the symbolic depths of life, and they carry life's numinous and primal energy. They have the power to create forms and shape events and to reconfigure our collective consciousness. They act as intermediaries between the visible, physical world and the inner realm of light, and when we recognize them as such, they align our consciousness with life's inner unfolding and allow us to participate consciously in the changes taking place at the heart of life. When we bring the divine spark of our higher consciousness to these new images of life's awakening, our spark speaks to the light at the core of creation and it helps the images come alive in this world with the full potential of life's highest consciousness and purpose.

But we will have to be awake to the real nature of what is awakening in order to recognize them. The images of the coming era will reflect life's oneness, its aliveness; they could emerge anywhere and will take unexpected forms.

The Internet is a striking example. The Internet is a powerful image of the changes happening at the core of creation, yet its deeper meaning and power have been veiled, partly by its very ubiquity. Carrying the energy of life's aliveness, it has quickly become so much a part of the fabric of our lives that we do not see its deeper symbolic dimension and the ways in which it is working to restructure our world. The Internet is everywhere, and for many of us it has become an indispensable tool of our daily lives. Yet it is still waiting to be recognized as one of the archetypal images of our future—a dynamic image of life's awakening to itself as a single living organism, rooted in the symbolic realm and carrying the real energy of life itself. The Internet not only reflects but enacts life's living oneness, its capacity to evolve and adapt, to quickly and continually reconfigure itself in an endlessly shifting web of connections.

But if it is to manifest its full potential, the deeper symbolic significance of the Internet needs to be recognized. It needs us to recognize its sacred dimension, its symbolic nature. Without this recognition, its higher purpose will be dimmed by the shadows of the ego-centered interests and desires the Internet currently serves, and the energy will flow in at a lower level, serving life's purpose less purely. The same holds true for all the emerging images of life. They need us to recognize their symbolic dimension. Our recognition can help life to reconfigure into the patterns, the archetypes and their this-world images, that will allow life to flourish again as a conscious expression and revelation of the Divine and Its all-embracing oneness.

There are no books, no schools or traditions, that will tell us exactly how to do this work with life's changing patterns. Only in the meeting of our consciousness with the awakening consciousness within life—our own awakened consciousness as we bring it to life in each moment—will we find the information we need. For at its core life knows everything that it needs to sustain itself; it knows how to evolve and how to recreate itself anew, and because it is a continual response to the divine moment, it knows precisely what is needed in each moment. It knows where our attention is needed, and how to direct it in the most beneficial way. It knows exactly what is required to bring our afflicted planet back into balance, to heal the Earth and the soul of humanity that has so painfully lost its way. And as it reawakens, this information will once again become available to us. Life is ready to teach us. Through this inner dialogue the intelligence within life can directly communicate with our intelligence and it will show us what we need to know.

What life knows about itself reaches far beyond the limits of our current understanding and the scope of our present sciences. When we bring our consciousness into dialogue with the world intelligence, we will open ourselves to new realms of knowledge and understanding, to a whole new science of life as yet undreamed of, to a whole new range of human possibilities. When we open ourselves fully to life as it really is, we will begin to experience again that life is alive, and that it can communicate with us directly and interact with us in many different dimensions—not just the physical dimension we now take to be the world, but also the symbolic realms, and the realm of pure consciousness and light, dimensions of magic and wonder and awe that have not been manifest in our world for a long time. When we once again allow life to express its profound intelligence, its playfulness and humor, its delight and its passion, we will find ourselves present in a world that is more dynamic and fluid and full of possibilities in each moment than we could ever believe possible within our current images of life.

JOY

The changes now taking place will bring a new light into the world, a light that is joyful and alive—and that will challenge us in ways that even the critical challenges of our present world do not. At every step we will encounter resistance, both within ourselves and within the world. For light is demanding. It forces us to face things as they really are. We might well prefer to stay in the shadows of all our desires and fears and inadequacies and all the dramas and conflicts they engender, where we do not have to recognize our real nature or the real nature of life. That recognition—that within our own consciousness we carry the light of life itself, that life itself is a living being that is coming alive now in a new way and needs us to help it, that the shape of the world to come rests upon the attention and participation we bring to it, that we are the guardians of life on Earth—confers upon us a greater responsibility than we have ever known.

But it is also a tremendous joy. This act of co-creation, this dialogue of our consciousness with the intelligence at the core of creation, is a living, creative relationship that is full of joy. Life as it really is may be demanding, but it is always joyous, because this is the nature of creation: an explosion of love that is continually coming into physical form. Being present in life as it really is means to reclaim joy as a central ingredient to life, an energy and quality of praise that is life's celebration of its divine source.

The light now awakening within life will awaken us to this joy. This is one of its purposes. The interaction of our consciousness with the light of life will remind us of what we have forgotten, that we are here to remember the sacred nature of life, and that the song of our remembrance is central to life. Our prayer and praise are always life's prayer and praise. We are a central note within creation, living the link of love between the Creator and the world that surrounds us. Without our note of praise and prayer, how can life evolve? The joy we will find in our real, co-creative relationship to life will attune us once again to that note of praise. Then life can reveal its secrets to us; it can show us again how to live.

The signs of the future point to the possibility of the spiritual awakening of the world itself.

THE AWAKENING *of the* WORLD
(2006)

Now bless thyself: thou met'st with things dying,
I with things new-born.

SHAKESPEARE, *The Winter's Tale*

SIGNS OF THE FUTURE

We are present at a time of transition when one era is dying and another is being born. At this moment we have a choice: we can stay with the images of the past—the ghosts of materialism that have polluted and desecrated our planet—or we can move into a future that is not yet defined, that is full of possibilities. There are already signs of this future, some visible and some as yet hidden. We can see the seeds of a global consciousness, a deep awareness that we are all one people who are a part of a living, interrelated ecosystem. And within this awareness is an awakening consciousness of oneness, a consciousness that is not based upon duality but upon a primal knowing of the oneness that belongs to life and is a direct expression of the Divine. In the different forms of global communication, we have been given some of the tools of this future, forms of interconnectivity that can take us beyond the hierarchical models of the past, into an organic and holistic way of living.

These are the signs of a future that is being born that can free us from the paradigm of duality and separation, which has caused so much conflict and division, the conflicts of masculine and feminine, the split between matter and spirit. What is awakening is a more integrated consciousness that can heal and transform many of the wounds of the past era and offer humanity the possibility of living in greater harmony with each other and the planet. This awakening consciousness carries an energy that comes from the source of life, an energy that is dynamically alive, has the joy of creation, and brings with it a more direct experience of the Divine that is within everything.

During the last era the many gods and goddesses of polytheism were replaced by the one God of monotheism. The masculine God of monotheism has a primarily tran- scendent nature—He lives in heaven—and as the gods and goddesses were banished, the created world was experi- enced as separate from its divine nature. The split between matter and spirit emphasized the spiritual poverty of the physical world, and much of the natural magic that belongs to creation was also banished, just as the wise women who practiced healing and other natural arts were persecuted. As we now become awake to the oneness that embraces all of creation and is an embodiment of the oneness of God, humanity can reclaim the divinity of matter and the wonder of divine presence. We no longer need to live in a world of separation which has divorced matter and our physical selves from its spiritual nature. Everything is a celebration of divine oneness, everything a unique opportunity to praise and remember God.

Every moment is dynamically alive with divine presence. Divine presence is not an isolated occurrence, not a single sighting to be revered and remembered, but an outpouring of love that is a constant stream coming into life. It cannot be captured, held as an icon. It needs to be embodied and fully lived. We are a part of the living substance of God that is in constant motion as it reveals its sacred nature again and again, "never in the same form twice." Each moment is complete, and each moment is a part of life's continual outpouring.

Once we awaken to divine presence we will find our self in a very different world. As we reconnect the physical world with its divine nature it can come alive again in many ways that are hidden or have been repressed. The physical world is not just dead matter, but a spinning organism of light full of magic and hidden potential. There are many forces in creation that are waiting to be activated that can heal and transform ourselves and the world. In the past some of these forces were known to shamans and healers. Much of the healing work of the future will be learning to work with these forces on both an individual and global level. We will learn new techniques that will bring together the wisdom of the shaman and priestess and the knowledge of the physician and scientist.

There are many different levels of reality, different dimensions of consciousness, from the physical world, through the archetypal world of symbols, to the pure consciousness of the Self and beyond into the planes of non-being. In the previous era the journey of spiritual ascent took the seeker from the physical world of the senses to the transcendent planes of reality. However, in the circle of divine oneness everything is present. There is no above and below, no ascending or descending. These are images of the past millennium. All the levels of reality interpenetrate, and the "highest" is present within each of us. The soul pervades the body from head to toe. Every cell is impregnated with divine essence. And between the particles of creation is the infinite emptiness, which recent physics suggests is full of dark energy, and mystics know to carry the secrets of non-being. ELLEN

The consciousness of oneness does not just embrace the physical world and the interrelationship of the web of life. It will also bring together the different levels of reality that in the last era were primarily kept separate. Our Western culture may have its roots in the Greek Pre-Socratic tradition of Parmenides and Empedocles, but we have systematically erased the knowledge that they were mystics and magicians with direct access to other worlds. Similarly we have forgotten how to work with the symbolic world, although as Carl Jung rediscovered, this knowledge was kept alive for many centuries by the alchemical tradition. We need to reclaim the knowing of how to work with the inner worlds, so that the wisdom and energy that are within us can come into the outer world and once again we can be nourished from the source of life and the meaning that comes from the soul.

Relearning the language of the inner world will mean regaining respect for the forces that underlie existence, the primal powers of life that were called gods and goddesses. We will have to realize the pain we have caused them through our neglect and abuse, and how our pursuit of materialism has not just polluted the physical world but also desecrated the inner world of the soul. We have created an inner and outer wasteland and stripped the sacred meaning from life. As we awaken to the new consciousness that is being offered we will have to accept responsibility for our actions and neglect. Consciousness always comes at a price and we will see more fully what we have done.

Then, when the worlds begin to flow together, when the inner and outer unite within our consciousness and our lives, something new can come alive. This is the child of the future, born from this *coniunctio*. She is both male and female, outer and inner, above and below, spirit and matter. She is within each of us and yet beyond us. She knows the secrets of matter and the magic within creation.

THE SONG OF THE WORLD

These are hints to what might come into being. But something even more fundamental is taking place, something that needs our attention and participation. The world is coming alive in a way that has not happened for many millennia. A light at the core of creation is awakening and speaking to the light within humanity. Just as we have forgotten that the world belongs to God, we have forgotten that it is a living spiritual being that, like an individual, can spiritually awaken. There was a time many eras ago when the world was awake and humanity was nourished directly from the Source. It is sometimes called the Golden Age.

What does it mean that the world might awaken, and how can we help in this process? Spiritual knowledge has always known about different levels of consciousness within the human being and developed practices to help us evolve and awaken to a "higher" level. An individual's awakening to the plane of the Self—a direct awareness of oneness, love, and pure consciousness—is the goal of many spiritual practices. The way the world functions as a living spiritual being is less known, although it does belong to certain spiritual traditions.[1] But there is an ancient teaching that says the individual is a microcosm of the whole, and thus what happens to the individual can happen to the whole.

The world as a living spiritual being has the potential for spiritual transformation. At this present time it has the possibility of waking up, of stepping into a new era of global consciousness. The new energy now present is a catalyst to help this to happen, to help the world awaken to a new awareness of its divine nature. But it needs the help of humanity to make this happen. We need to reclaim our ancient role as guardians of the Earth and consciously participate in the transformation of the Earth. But in order to do this work we need to leave behind many of our patterns that focus solely on our material well-being.

Life has always presented humanity with new challenges to help us evolve and grow. Yet at each stage we are reluctant to "take upon us the mystery of things," but would rather remain with the attitudes and conditioning that belong to a previous era. But a new way of life is waiting to come into being, and it needs us as midwives. We just need to recognize the simple wonder of being alive, of being a part of life, and say an unconditional "yes" to being fully present at this moment of transition. Through this "yes" we open ourselves to what life needs. Then life can reveal to us how we can work together. Life is a living spiritual presence that knows its divine purpose of which we are a part.

This is the challenge that is confronting those who are drawn to this light, whose souls want to work with this awakening. Here there are no righteous wars or the dramas of good and bad. Here there is no one and nothing to be saved or redeemed. A new birth has no knowledge of such things. It just carries the primal spark of creation, the eternal yes of life coming into being, and the stamp of the real Creator.

What could it mean if the world awakens? This is like the wonder of our own awakening. Over many years we might be given glimpses of the power of the heart, the wisdom and love within. We work to purify ourselves, to contribute in service to life and love. And one day something comes alive within us. Our heart awakens, and we experience a greater power and knowing beyond ourselves. This is the potential of the moment in the world. Our world's soul can awaken, and all humanity can come to know its true nature.

We know how our awakening frees us from many of the patterns that restrict and deny us our divine nature: how in the light of our real Self many of the problems that consumed our energy fall away, or fade like mist in the morning sun. Although life presents us with difficulties and challenges, it is not a problem, but a living spiritual being made in the image of God. When we recognize life's divine nature, not as a spiritual ideal but as a simple presence that is part of our everyday existence, we will find that once again we can experience the simple wonder and joy that belong to life.

When we work together with life as a living divine wholeness we will see the simple answers life has for sustaining itself. And we will be given new energies to work with life, energies that are presently hidden within creation but are also waiting to come alive. And something else will be present, something that we have forgotten for many millennia. In the midst of creation there is a song, a song that belongs to life that tells of its hidden purpose, what the Sufis call the mystery of the word "*Kun!*" ("Be!"). Seeing life through the eyes of the ego we only see a small fraction of what it means to be alive, but in this song of life its true wonder and meaning can be heard. To be the guardians of the planet, to step into our real role of global responsibility, will open us to this song, this mystery made manifest.

An awareness of global oneness has begun to constellate. The idea of the unity of life, that "we are one," no longer belongs just to a spiritual or ecological fringe, but is becoming part of the mainstream. But this awareness is lacking an essential ingredient—it is still an idea, it is not fully alive. When it becomes alive, the heart of the world will open and we will hear the song of the divine oneness of life.

The world has to awaken from its sleep of forgetfulness—it can no longer afford to forget its divinity. More than any pollution, it is this forgetfulness that is killing the Earth. Collectively we are dying—we have forgotten our purpose, and a life-form that has forgotten its purpose cannot survive. Its fundamental reason for existence fades away. The awakening of the heart of the world can redeem what has been desecrated, heal what has been wounded, purify what has been polluted. The song of the world will remind us all why we are here and the whole of life will rejoice.

Life itself is bringing us together in different ways, recreating an organic pattern in the emerging network of human connections.

A RETURN *to*
WHAT IS SIMPLE
(2006)

Life itself is a continual force of change, always creating new forms. We have the choice to respond to this dynamic of change, or to stay with the forms of the past. This is especially true at this present time. If we are to embrace an entirely new way of being and living together, we must let go of many patterns of living and ways of relating. But do we recognize the changes that are possible? Are we trying to work with them, or do we fight against them? Are we prepared to embrace the new forms that life is offering? Or are we still caught in the old images; for example, the sense of well-being and security that comes from material prosperity?

It is time to look closely at what is really happening. The mystic has always known that in order to find the cause of any effect we have to turn our attention inward—to look at the inner patterns. We look to the hints in our dreams, we read the images of our psyche that are not censored by our conscious conditioning. When Joseph interpreted the dreams of the seven years of plenty and the seven lean years, Egypt was saved a catastrophe.

And yet we have rejected the images of the inner as belonging to a mythological past or the psychiatrist's couch. Instead, we try to listen to the voices of the outer experts. But with so many newscasters, political and economic analysts, and even spiritual visionaries, how do we know who and what to trust? And do we even know how to listen?

But if we look carefully we can find a thread that links it all together—links our dreams and the stories on the news, links the trivial, the mundane, and the sensational. There is a thread that is our collective destiny, and it is inside each of us as well as in the world around.

This thread is so simple it is overlooked. It is so ordinary we pass it by. It is in our hope, in our need to be loved, in the warmth of a handshake or the touch of a kiss. It is in the most basic connection between human beings, not the words we say but the very nature of communication. It is in the simple fact that we are all connected together. It is in the primal knowing that we are one.

Because we live at the end of an era, life has apparently become more complex. This is one of the signs of things falling apart. With our computer-generated models we look for complex answers to our problems.

But the signs of the emerging culture are not complex; they are in patterns that unify, that bring things together, so that they do not fall apart into a myriad pieces. The danger comes when we turn away from what is offered, through either ignorance or arrogance—when we stick to our models of ever-increasing complexity rather than recognize the simple human values that belong to our being.

We do not have to save or protect our culture. We do not have the power to resist the dynamics of change. Nor do we have to create a new culture. We have neither the energy nor the knowledge for such an undertaking.

But we do have a responsibility: to listen, to love and be loved, and to become aware of what is really happening. We have to accept that *we* cannot save the planet, just as we cannot defeat the forces of corruption. Enough battles have been fought, and the planet is a living being that can heal itself, with our love and cooperation.

What we always seem to overlook is the simple wonder of being human, which means to be divine. We are the meeting of the two worlds, the place where miracles can happen and the Divine come alive in a new way. We are the light at the end of the tunnel. We are the warmth and the care and the compassion, as much as we carry the scars of our cruelty and anger.

This coming change is so fundamental, it is a return to what is simple and essential, what is basic to life and yet not easy to live. There are forces at work that push us outwards towards complexity. These are the forces that take away our joy and demand that we work harder and harder. They drive us into struggles and conflicts we do not need, and always try to obscure the simple joy of life, of being together and valuing our companionship. Fast-food and mega-movies may glitter and catch our collective attention, but we know in our hearts that something fundamental is being over-looked. We do not need to drown in prosperity. Nor do we need to impose any beliefs on others. We have simply to recognize what is *most meaningful* and live this in our own lives. What is meaningful has the energy and light to free us from so many imposed beliefs; for example, the belief in consumerism that feeds the greed that is destroying our planet.

In the simplicity of our human values—love, and joy, and hope—we are all connected together. But we can only discover this connection when we return to this simple core of being. Otherwise we will fall apart with a world that has lost its center, a world that believes in its own advertising slogans. When we return to this connection of the heart we will see what is being born, how a linking together of individuals, groups, and communities is taking place, how patterns of relationships are growing—and how life energy is flowing along these patterns. Once again humanity is recreating itself, creating a new civilization amidst the old.

In our focus on complexity and the technological gadgets of our culture, we have overlooked the rule that the more complex something becomes the more its energy becomes scattered and easily fragmented. Human beings have a unique role as the microcosm of the whole, which means that they can carry the whole multiplicity of creation within the simplicity of their natural self. In this simplicity we also carry the primal power that belongs to life and to our divine nature.

When human beings are not scattered in the "ten thousand things," we are very powerful. This is a part of the purpose of spiritual practice: as we return to our essence we become more focused and able to claim our own power. In the simple core of our being we carry the imprint of the Divine and its miraculous nature. This quality of life is able to express itself more directly in our love and joy, and other simple qualities: for example, in the way we come together as groups and communities and support each other. Often this is more evident in times of disaster and real need, when we instinctively respond to the primal bond between us. Then help and life energy flow through these ordinary human connections. And in this energy and flow is the simple power of the Divine within life, going where it is needed. How often is the help from neighbors, friends, and even strangers in these times more valuable than external structures and organizations.

If we look carefully at this time we can see how life is regenerating itself in the most simple, ordinary ways. The signs of this regeneration are all around us, how people are coming together in different ways. Social media are part of this process in the way they connect people together regardless of the barriers of race or nationality or physical location. Different people in all parts of the world are linking together, forming networks of shared interests. These networks are outside the control of any hierarchy or government. They belong to life itself.

Patterns are growing that link people together in new and diverse ways. We have yet to fully understand that it is these patterns of relationship that are so essential, that provide the simple answers to the complexity of the times. They are not just for conveying information, but are creating a new, fast-changing, organic interrelationship of individuals and groups. Something is coming alive in a new way.

People are making connections on many different levels, through global trade, travel, telecommunications, conferences and other forms of gatherings. Interfaith dialogues are one level of interreligious communication. The migration of spiritual paths and traditions from the East to the West has also formed a level of global connection in which East and West are merging, creating a spiritual light that is "neither of the East nor of the West."

What we have yet to realize is that this is all a part of the organism of life recreating itself on the pattern of oneness. We see these changes with the eyes of individuality and fragmentation that focus on the individual parts. We are still caught in the complex images of a decaying culture. The real picture is an emerging wholeness that is a life force in itself. *Life is reconnecting itself in order to survive and evolve.*

In these patterns of reconnection a new life force is flowing. This life force has the urgency that is needed to survive and change at this time of crisis. It also has the power of oneness, and the simplicity of bringing people together. It is about sharing rather than possessiveness and isolation. It is the deep joy of knowing that we are one life.

During the last century, deep within the soul of the world, shifts have been taking place that prefigure these present changes. And yet as the darkness tries to deny us any real change, the world needs our prayers to help it to transform.

A PRAYER *for the* WORLD
(2008)

Spiritual teachings have long told us that changes happen first in the inner planes and then gradually become manifest in the outer world of our daily life. Today we see signs of a world changing more and more rapidly, spinning out of balance, suggesting that a shift is taking place. If one wants to explore and understand what these changes might mean, one needs to go within, to where the forces that define our surface life constellate. My own journey has drawn me into the chambers of the heart, into the love and emptiness that are the home of the mystic. But this journey has also had a global dimension. I have been taken into the worlds that underlie creation, where I have been shown how the energies in these inner worlds are shifting, and how at this time a new energy is constellating, new patterns are evolving.

This exploration of the forces underlying creation began over thirty years ago with a series of journeys in the archetypal world, a reality long known to the shaman. Carl Jung describes these archetypal forces as the great determining factors within humanity, their patterns the "riverbeds" through which the waters of life flow. And as I journeyed in this ancient world I experienced how certain archetypal patterns were changing, certain new forces were awakening. When these forces begin to shift in the depths, one knows that something in the core of humanity is changing.

During these journeys the archetypes taught me much about the patterns behind existence, and how their own power worked within this ancient order. For example, one figure, the Ancient of Days, showed me the meaning of time and how the growth of plants, the cycles of our body, the tides of our seas, are all reflected in the movement of the stars and the galaxies—the traditional understanding of "as above so below"—and how we have lost this flow and its purpose in our hectic lives. Instead we fight with time, trying to cram as much as we can into our hours and days, rather than revering its deeper meaning.[1] And so we have wounded the archetype of time, even tried to imprison it within our schedules.[2] In turn it has turned bitter, giving a bleakness to the passage of our days.

Another figure, whose body was made of plants and fruits and all growing things, showed me the endless bounty of the Earth, the generosity of creation. And yet because we have not revered its sacred nature, we have lost the joy that is essential to life, the joy that is so evident in the moment a flower opens, or a fish's scales are caught in the sunlight as it swims through the water. As a result we work so hard for so little, and have denied our self and our children the music and colors of creation, life's essential giving and generosity.

As much as I was shown their magic and knowledge, I also saw how these ancient archetypes—the forces that used to be revered as gods and that have underlain our existence for so long—were rearranging themselves, making way for a new figure. This new energy would change the patterns of existence in ways beyond even their vast understanding. They knew that their time of power was coming to an end and they welcomed it. They all had been wounded by the way our rational culture had disregarded their existence, the way our surface life had lost its conscious connection with the depths. They understood that unless this new energy could come into existence, something essential within the world would die, and even with all their power and wisdom they would not be able to stop this death.

As these forces began to shift, old patterns of power began to fall away and a new archetype emerged. This archetype appeared to me in the image of the cosmic child with stars in its eyes. And this child described itself as a space rather than a form, a sacred space in which we can be our true selves and experience the joy that belongs to life itself. This space includes everything, each according to its true nature: "all can be themselves as they are, as they were in the beginning." This child also brings together the inner and the outer, spirit and matter, uniting the opposites in oneness.

This child is not born of the past, not even of the ancient order of the archetypes, but has come from a totally different dimension within our self and within the cosmos. It carries a possibility that is free of many of the patterns of the past, and it also has a magic that enables things to happen in a way they have never happened before. It is alive in a completely new way, belonging to the moment that is outside of time and yet fully present. It has similarities to the Christ Child, born in the simplicity of the manger, amidst the ordinariness of life, and carrying a new possibility of love. But this archetype that is appearing has a cosmic dimension that is rarely present in Christian symbolism. This cosmic child is an unexpected promise and gift, and comes with laughter and joy as well as being a messenger from far distant galaxies.

Through the eyes of the child I began to become aware of how much we have restricted ourselves, imprisoned our human potential in the forms of our rational conditioning. And now coming alive within the depths of humanity is a new way to be, a way that returns us to the sacred core of our being and its magical potential. If we have access to this archetypal energy, many of the toys of triviality that we use to distract and entertain ourselves will no longer be needed. Once again we can be nourished directly by life and its inherent joy. And in this reconnection are many possibilities of healing both ourselves and our planet. We would again be given access to the magic within creation.

At the beginning of the year 2000, I began to see this new archetype as a spark of consciousness waiting to be given to humanity both individually and as a whole. This consciousness of oneness is well-known to the mystic, because it belongs to the consciousness of our Higher Self, or *atman*, which exists in a plane of love and oneness, in which everything is included and known according to its true nature. It is symbolized by the mandala in which all the different parts form an image of beauty and wholeness. But this new consciousness is not just a spiritual ideal, reached through years of meditation, but a tangible reality already visible in our global communication. It is present in the Internet and cell-phone networks that form an interconnected whole, a dynamic oneness to which more and more of the world has access, from the cyber cafés in Uzbekistan to the cell phones of Somalia. And yet claiming this consciousness also requires that we make a step away from our present ego-centered focus and take responsibility for the whole, as our ecological crisis is demanding.

After journeying into the depths I began to see the darkness that surrounds us now, and how it does not want something so precious as this new consciousness to be given so freely. I saw how we are controlled by the forces that manipulate us into greed, that create the web of desire that holds us in the grip of this consumer society that is poisoning our souls and the planet. These forces keep us focused on our self, our own wants and needs, and also busy on the surface, away from the inner energy that might free us. They can be seen at work within the multinational companies that control so much of the buying and selling we call life, and yet they come from a deeper place within creation, a place that does not want any change, that does not want the light. These forces are also ancient and power-ful; they know cunning and deceit and how to make us sell our souls for a few pieces of silver. They offer us the image of progress even while they deny us the sacred light that alone can nourish our souls and our daily life. And they are feeding off the lifeblood of the planet.

Not only do they draw us into this web of wanting, but they also make sure that we remain unfulfilled, addicted to the products that promise so much and give so little. And these forces are even more dangerous because they are covering the world in forgetfulness, the forgetfulness of our divine nature. How can we reconnect with our soul when we have forgotten we have a soul? How can we reclaim our divine nature when we have forgotten we are divine? The forces of forgetfulness are very powerful and we are so easily seduced by the false sweetness we are offered. We become indentured servants selling their products, not even know-ing the light we have left behind.

And then I was taken into the very core of creation and saw that the shift that is being offered to humanity and to the whole world comes only once in thousands of years. This is not just the shift into a new age, but into a way of being that has not been offered for many millennia. And the world is included in a way that it has not been since the beginning, since the early days when consciousness was first present within humanity and everything was sacred. This time is just a distant memory in our collective consciousness, a time "before the fall" that we retain as an image called paradise, when in the imagery of the Bible we walked in the Garden of Eden together with God. This was a time before time, when humanity was not separate from the sacred, when the simplicity and directness of our divine nature were present in every breath, in every step. Sometimes, even today, one can encounter children who still live for a few years in this magical world, before the coldness of our competitive culture freezes them out of its warmth. But I was shown that humanity has the possibility to claim this connection once again, to no longer live in separation, to no longer live in exile.

And then I was reminded that there was a time when the heart of the world was awake and sang the song of creation. And I heard this ancient song that brought tears to my own heart, and I knew that if the heart of the world started to sing, all of creation would come alive in a new way, that it would be like spring at the end of thousands of years of winter. And I saw how the sacred colors would return, and birds of remembrance would sing, and how we could reconnect with all the different creatures of the Earth and know their language and their ways and once again live in harmony with all of creation. And the doors between the worlds would open and all levels of existence could commune and nourish each other. Even the presence of angels and *devas* would no longer be a supposition but part of the lives of all the inhabitants of this world, as it was in the long distant past.

On the most practical level, as we reclaim our connection with the inner worlds, their many inhabitants can help us. The nature spirits can teach us how to live in harmony with the forces of nature, how to grow crops that are sustainable and nourishing. There are also great *devas* that can help in purifying the pollution in the air and the water, as well as teaching us once again the healing properties of plants and the magic of stones. It is time for us to acknowledge our kinship with creation, and with the forces within creation, to learn how to communicate with the elemental inhabitants of the planet, to respect their ways as well as gain from their knowledge. We are all part of one living whole, each part contributing. The *devas* and other spirits will also gain from communion with humanity, for they too have suffered from this isolation, often withdrawing deep into the inner worlds as we have denied their existence, polluted their sacred streams, cut down their groves.

In the West we have forgotten much about the inner worlds, though in recent years there has been some increased interest in nature spirits and how to work with them,[3] and we have kept an awareness of angels. But it is so long since the heart of the world has sung that we have even forgotten its existence in our mythological memory. There are traces of it in the Australian Aboriginal songlines through which the Aboriginal people find their way through the bush. However, we do not even know that the world has a heart, let alone that it can sing. But there is the possibility for this song to return, for the world to awaken again. When I felt the promise of this primal music it stirred something deep within me that belongs to the magic of the very beginning, when everything in creation was given its sacred name and purpose.[4] When we feel this song we will know what it really means to be alive and to be a part of the many dimensions that are present in existence.

And I saw the new knowledge that was waiting to be given to the world, the technologies of the future as revolutionary as the Internet has been in the last few short years. I knew that the understanding of the power of light would be primary, because sunlight is free and can be used in a way that does not cause pollution. It is also symbolically time for us to connect with the power source at the center of our solar system, and to bring that power down into our daily life. The same power that ripens our grains and fruits can meet our energy needs. Also our understanding of the spiritual light of a human being will evolve and become accessible in ways that are hidden from us at present.[5] We will be given back more of our spiritual heritage, the Divine that is waiting within us to be lived.

The ways of the future will be simple rather than complex, and they will be given to humanity as a whole, because wholeness is the stamp of the coming era. The time of the warring brothers is past and oneness is beginning, and the ways of oneness will always be simple. Yes, there will be a dark side to this coming era, a misuse of its magic and powers. On this plane of existence there has to remain the duality of light and dark, and human beings will be tested as they have always been tested. But no one will want to return to the grey world of our present era, with its many divisions and deprivations.

So much was shown to me in the inner worlds, the light and the darkness. And I saw the places of power that belong to the future, waiting to be awakened, waiting for their energy to be activated. The power of places like Stonehenge lies in eras of the distant past, but there are new places waiting whose energy belongs to the next step in our human evolution, places that will bring together the energies of heaven and Earth in a new way, and use the sacred power that belongs to both. And there are ways to activate these places, no longer with buildings, pyramids, or stone circles that mark the rising of the sun, but with small groups of initiates who know the words to be spoken, the prayers to be said, who hold in their hearts the key to these places and will be given the wisdom of how to use their power for the benefit of humanity.

In many ways it is small groups who will become the centers of power of the new age, forming a network of light. Small groups of mystics and seekers are already holding a new light for humanity, forming a web of light around the world. They have no structure or organization, do not even know how they are linked together. But they are part of the organic structure of the light body of the planet, helping in its evolution, very different from the hierarchical, organized forces of darkness that are so visible. And I saw how a thread of divine awakening was being woven into the light body of the planet, and how this could help in its healing. And how the darkness was always trying to divert our attention, to take us back to the ego and the curse of consumerism.

And so the drama of our world continued to be played out through the weeks and months and years; and the world was dying, and the soul of the world was crying for help. And some souls were responding to this call and others were caught in forgetfulness. And the heart of the world was waiting.

On the world stage there were wars and the fears of terrorists, a stock market bubble and a real estate and banking collapse. And all the time within creation a light was trying to be born, and some souls were drawn to act as midwives, to help it to be born. I saw in my prayers and meditations that there was a new note in the love at the core of creation, and that this note of love had not been played before. This new note of love is like a key that opens a new level of perception within us, awakens centers of power within the world and within humanity.[6]

And I saw how the darkness was gathering its forces to make sure that nothing new could be born. As always the destiny of humanity in this world is played out between the forces of light and darkness, but at this time of transition this contest of forces is emphasized. While the light and the love struggle to give us our divine heritage, our freedom and joy, the darkness finds new ways to distract us and to deny the light its power. And I saw more and more clearly that alone humanity could not make this transition. It has been too long since it knew about the workings of the inner worlds. Its shamans and seers have almost all gone. We have forgotten the places of power and the sacred words. Instead we are left with a few new-age teachings, which may bring us some light but have little understanding of the forces in creation.

And so we are left waiting in the gathering darkness. Slowly the light of the last age is going out. Slowly the meaning of life is being lost, until, as in Shakespeare's *Macbeth*,

> Life's but a walking shadow....
> It is a tale
> Told by an idiot, full of sound and fury,
> Signifying nothing.[7]

The light of the sacred is something fundamental to human existence: only the sacred gives real meaning to our everyday lives. And yet our rational culture has rejected it, dismissed its symbols as superstition, so that we no longer even know what we have lost. And not content with polluting the outer world, we have taken our greed into the inner world, using its symbols to advertise and sell products we do not need. We have systematically prostituted the symbolic world for corporate and personal profit. Like Macbeth we are killing our king, our sacred self, for the values of the ego, for our own power and gain, not really knowing what this means until the light goes out, and we are left with only the wasteland of the inner and outer worlds.

Just as a light is waiting to be born within the world, the light as we know it is dying, is fading away. This life without light, without meaning or purpose, is what we have created or been drawn into. And there is a moment in cosmic time when the soul of the world can no longer bear the absence of light and will forget its own divine purpose, will become lost. When the light of an individual soul is lost, it cannot find its way. It becomes a lost soul, wandering in the darkness of forgetfulness without purpose: "And to whomsoever God assigns no light, no light has he."[8] And it is said that what happens to an individual could happen to the world soul, to the light within the world. Then the light in the world could not be reborn, but would remain extinguished, and the dark ages will come upon us.

The time when one light fades before another light is born is always the most dangerous. It is a time of transition when the fate of the world hangs in balance. Will we be the midwives of a new consciousness, giving birth to an era of oneness? Or will the darkness come upon us, denying us any purpose, making it impossible for us to find our way? And sadly we seem ignorant of this moment, as if our patterns of denial are so strong we are unable to take responsibility for our fate or the fate of the world. The signs of the need for change are all around us. There is even a deep anxiety that is beginning to surface within our collective Western consciousness that life as we know it cannot continue. Forces are building up in the inner and outer worlds. Those who listen to their dreams and the dreams of others have seen the tsunami coming, the dark clouds gathering on the horizon, as well as images of a new future, a return to the sacred and the simple joy of life. And yet collectively we have been conditioned to think that our problems and their solution are only in the physical world around us, and so we are denied access to the inner light and help we so desperately need.

There is an ancient prophecy that at a time of real crisis, not just a physical disaster but a crisis on the level of the soul, divine intercession is possible.[9] Many people have experienced this within their own life, how a time of despair and darkness becomes a moment of grace. But we have forgotten that what happens to an individual could also happen to the whole world. We have mythic memories of the flood when an angry God sought to destroy the darkness, but we do not know what it might mean for the power of Divine Light to intercede. We have forgotten about the power of God. It is not even present in our prophecies.

We have done so much damage to ourselves and to the world, and the forces of darkness have hidden us from the light. We are caught more deeply in the darkness than we know, but this is one of the powers of darkness—it does not let us see the danger. From my own inner experience I know that only the Divine can effect real transformation, and that Divine Presence is needed to heal the world. As a mystic I know what it means for a fragment of Divine Power to come into my heart and into my life: how It can change everything in a moment. I know the absolute reality of This Presence and how the forces of darkness cannot withstand It.

On our own we cannot make the shift that we need to make, and we can no longer afford to wait. Soon all the light will be gone and the possibilities for the future will be covered over. From the hearts of those who belong to God a prayer for the world is born, an impassioned cry that echoes the cry of the world soul to call the Divine back into this world, to open the door of our hearts and the whole world for our Creator and Beloved. There will be a price to pay for our forgetfulness, as we finally take responsibility for our ego-centered culture that has ravaged the world. Maybe this is one of the reasons why we hide from the light. But how can the world be reborn without the presence of Divine Light?

My personal prayer is: May the Beloved return and take this world in both hands, and with tears and tenderness wash away the debris, with power and light pierce through the darkness and awaken the light that is at the core of creation. This meeting of light and light will spin the world on its new axis of love, its new place in the unfolding wonder of the cosmos. Without the return of Divine Light our world will drift without meaning, lost in the backwaters of existence. We need Divine Presence more than we know, more than we dare to acknowledge. We have failed in our guardianship of the planet, forgetting its sacred substance and purpose, using it merely for our own gain. We have forgotten to read the signs of God. And now we are living in a moment in its history which will determine its future for many millennia. The world knows that it is dying, and although we cannot bring it back to life, we can turn our attention to the One Being whose power and beauty can.

We are the prayer of the world, our cry is its cry. We carry within us, within each of us, the knowing of our divine origin, and our hearts each have their own connection with what is most sacred. We need this connection of love to be the messenger to our Beloved, to remind the Divine of our weakness and need, of our failing and forgetfulness. Each in our own way we make our prayer, we turn towards love. And this turning of our hearts is our deepest offering, the thing that is most real in this world. We pray that our Beloved hears this cry in our hearts and the heart of the world. And we know that all is according to Divine Will.

In this time of darkness before the dawn we can only turn towards the One.

WAITING *for the* DAWN
(2009)

Whatever names we use to describe our hope for the future, for real change in our world, we are together waiting for a dawn, for a new light. But we have been standing on the edge of this dawn for so long now, our souls dreaming of its coming, that when the dawn does finally come, will we notice it? There is a danger that our eyes have become so accustomed to the present darkness—to its attractions and distortions—that the dawn could easily pass us by; that our patterns of avoidance are so entrenched, our pursuit of self-centered pleasures or problems so pervasive, that we will not be able to see something as simple as sunlight. Or we will see something, but because our attention has been for so long in the half-light and shadows, we will pass it over, presuming it is just another mirage, another false dawn.

Personally I no longer know what to wait for. As I have watched in the darkness my own dreams have been discarded, torn up and thrown away. Maybe I am just another soothsayer speaking prophecies that are blown away in the wind. Unlike prayer flags they draw no attention either in this world or the next. I have waited for this dawn for far too long. I have seen and spoken of our present desolation, our forgetfulness and failures, until my words have fallen into silence. And now? Could a coming dawn be real? Could the sunlight ever return? Or is it all just another empty promise of which we have heard so many? Maybe laughter is all that we are left with, a laughter at the futility of any dream, because what is Real is so other, so completely different, so unavoidably present.

Yet in the depths of our hearts, in the recesses of each of our souls, we long for this dawn. We know, even if we pretend to our self and others to have dismissed it, that we cannot live without it. We are starving, destitute in ways we no longer are able to articulate. The symbols that used to nourish our souls have all gone and only slogans remain. Our desolation has been present for so long we pretend that it is normal, acceptable, and so we cover it with desires and fantasies—not the dreams of our real imaginings, but fantasies that are best forgotten. And each day our soul and the soul of the world is crying, and we do not dare to hear it.

And yet this forgotten and longed-for dawn is coming, either within our hearts or within the world or both.

Otherwise the darkness and the endless distortions it creates would be all we have left. Otherwise our breath would no longer carry the possibility of remembrance, or a return to the Source that is our real Home. Otherwise ... It is too terrible to think of a cycle that has stopped turning, that remains fixed in this darkness, in this polluted and fractured landscape we have created in the inner and outer worlds.

But what is this dawn we hope for and have forgotten? Will it come with a cry of heralds, with angels of light and glory, with devastation and blazing light? Will it arrive with the soft caress of a lover's touch or the kindness of a mother? Will it creep into our consciousness, unannounced, as our attention is attracted elsewhere? Will we even be allowed to see it, or have our failings been so great we are to be veiled from our own and the world's redemption, like Moses not allowed to walk in the promised land? It cannot depend upon us—we have passed over so many opportunities; again and again we were distracted when the moment to participate was present. We have avoided our real responsibility. But is it possible that it is not about us at all, despite our hubris which makes us think that we are the center of every event?

It will come in its own way. The dawn will belong to the day that belongs to God, to that secret that is at the core of everything that exists. How can it be otherwise? A dawn that belongs to humanity is not worth waiting for, crying for. The dawn must belong to that Being we call Divine. It is not our cycle of existence, however much we think that we are the center around which the sun moves daily. This world with all of its chaos and beauty is something so divine we would tremble if we could glimpse it. And it is this Divine that is the dawn that is coming. This is the song for which our hearts are waiting, the song whose chorus we may be allowed to sing together with all of creation.

Yes, we have missed every opportunity. Yes, our arrogance has claimed that it is our world to rape and pillage as we like. We have done so many misdeeds. But how significant are these failings compared to the wonder and mercy of that Eternal Being? Only that One Being is real, and we have suffered from the illusion of separation for too long—it is time to remember and return to the oneness from which our myths tell us we were banished. It is time to hang up this cloak of separation that has caused the hubris of our contemporary ideology, our fixation upon ourselves. The dawn is a return to oneness as much as it is a bowing down before God, a recognition of a sunlight that is real and cannot be covered even by the clouds of our ego-selves.

There will be a price to pay; there is always a price to pay for forgetfulness such as ours. We cannot dismiss the Divine and allow our greed to ravage so much of Its world without there being some retribution. We are no longer children living without responsibility or consequences. But this should not be our central consideration. We should be grateful to be allowed to be present when the dawn comes, to be a part of this dawn that is breaking—because the mystery of oneness means that like everything else we belong to this dawn, even if we have the free will to turn away. Our real need is as always to be awake, to be attentive, to be present as the light comes into our hearts and into the world. It may come so dramatically we cannot ignore it, or it may come so quietly, so hiddenly, it will steal away the darkness like a thief we do not notice until afterwards. Or it may be something as simple as birth, where pain is a passage into joy and love. All that matters is that our eyes are open and we are present at this moment, and hopefully, that our hearts and minds are turned towards the One.

In the previous chapter I wrote about the coming dawn and our need to be fully present to welcome the sunrise. But I also wrote about the darkness before the dawn, and it is this darkness that I am now drawn to articulate more fully, to explore its meaning and the story it is telling us.

THE DARKNESS BEFORE *the* DAWN
(2009)

We are standing on the edge of an abyss, this crisis of climate change and ecological devastation. And echoing this physical tragedy is our deep forgetfulness of the sacredness of creation—an attitude that has allowed us to treat the planet as an object we can pollute and desecrate at our own will. What has drawn us into this inner and outer self-destructive cycle? Is it just greed, the desires of materialism that consume so much with so little regard for their source or their real cost? Or are there other forces at work?

It is easy to see the ravages of our materialistic culture, a society that puts short-term economic "progress" above any larger or more far-sighted concerns. But is this surface-self only to blame, or are its attitudes and their devastating effect part of a larger picture or pattern, a deeper dying to what holds us all together as a sacred whole?

Psychology tells us that our personal self and ego-self are surface identities under which there are other, often more potent forces: our repressed selves, and the deeper forces of the archetypal or collective world. We know how our repressed self can easily erupt in anger or catch us in neurosis or other dramas. We can begin to glimpse the powerful inner forces that express themselves in the images of dreams or myths, and recognize how these belong to our collective heritage and destiny. But what is the real story being told at this present time? What is our surface drama, with its economic and political concerns, and what are the deeper forces at work and the story that they are telling? Who or what is calling us to this abyss that we appear to approach so willingly?

Poets and prophets have warned us of this time, when, in Yeats' words:

> Things fall apart; the centre cannot hold ...
> Surely some revelation is at hand;
> Surely the Second Coming is at hand.[1]

But is it a second coming we are approaching or a dark age? Is this the longed-for golden era preceded by a global crisis intense enough to awaken us, or are we becoming more and more lost in a nightmare of forgetfulness played out on our world stage when the forces of life turn against us? Are our ecological destruction, the dangers of a global economic collapse, the forces of terrorism, and the loss of the sacred in everyday life all part of a pattern, which, like the destructive reality of the suicide bomber, is now here to haunt us? And what might this mean to our souls as well as to our daily outer life?

The danger is always that we content ourselves with the surface dramas of the moment and our desire to solve the problems they present without concern or understanding of the underlying causes. But can our small light either confront or dispel those deeper shadows if we do not even acknowledge their existence? Can we afford to neglect this inner reality just as we have neglected the fragile state of our ecosystem until it is almost too late? Do we not know that it is all an interconnected whole, that the state of our planet and our loss of the sacred is one story told in different ways? Is it not time to listen to this story, to find its meaning, and to recognize finally that this story belongs to each of us—it is our own story as well as the story of the whole world?

Many of us are awakening to a desperate need to save our planet. But if we do not recognize the real nature of the forces we are encountering, how can we work to redeem the present situation? Is it the monster of corporate greed and exploitation, the forces of fundamentalism, or simple ignorance and self-centeredness? Or is this an archetypal drama in which we are unknowingly caught—the dark myth of material prosperity whose empty promises are that it will fulfill us, bring us the happiness or security we crave, while the truth is that it starves our souls and robs real meaning and joy from our lives?

This dark side of our addiction to consumerism is one possible explanation of what is happening. But how much do we understand the power of this inner darkness in our everyday life? Are we unknowingly victims of primal forces, gods we have long dismissed but now desperately need to acknowledge and honor? All we know for certain is that for the first time we are bound together in a global drama from which we cannot isolate ourselves. This is the most basic truth of global pollution and climate change, as well as the threat of terrorism. There is no "over there" anymore.

Spiritual teachings tell us that what happens to the individual affects the whole. Now we know for certain that what happens to the whole affects each of us. Unless we remain buried in an illusion of insularity, we have to acknowledge that we are all interconnected. This was one of the lessons of the 2008 financial crisis, the "sub-prime mortgage crisis"—how greed and bad lending practices in some parts of America almost brought down the whole global financial system. And its resulting recession has affected us all. These patterns of interconnectedness are the first major lesson of this crisis.

But what of this darkness that we label individual or corporate greed? We may seek to condemn the bankers with their million-dollar bonuses, but are they the cause or just a symptom of this darkness that appears to disregard everything except immediate profit? This ego-centered profit motive is at the very center of our Western ideology. It may have promised us prosperity, but there is a darker cost we are now beginning to pay. Could it be that through our human weaknesses of ignorance and greed a deeper darkness has entered and spread its tentacles, slowly sucking the sacred lifeblood of the planet, leaving us in this physical and soul-less wasteland we now inhabit? A darkness that is gradu-ally drawing the light and meaning from our lives? It takes away hope and leaves instead anxiety, fear, even an anger that we may project onto outer situations or events, but that comes instead from deep within, from empty promises or a sense of betrayal.

What is the purpose of this darkness? Where has it come from and where is it taking us? What is its story? Does it just feed off the life and light of the planet and ourselves, until all that is left is a soulless wasteland—an inner and outer landscape whose sacred streams are polluted, ancient groves cut down?

This darkness is all around us, even if we try not to notice it. We may read about oil spills and see its images on our TV screens without noticing the affect on our souls. But knowingly or unknowingly we have become the story that this darkness is telling. We are the light that is being lost. We may blame corporate executives or politicians, decry the evils of terrorists, but this is no absolution. Once we dare to look closely both around us and beneath the surface we can sense that here is a deeper darkness that can't be caused by any individual or group. We can feel it even if we cannot name it.

. The darkness is as real as the pollution and more dangerous because we do not know its source. We do not know why it is here or what it wants. We have not dared to see it because then we would realize how we are all implicated. Even if we look to the light we have to acknowledge the darkness that is within and around us. And if we just want to follow the desires of the ego, we need to know the danger. Ignorance cannot be an excuse.

And most important we need to see how the darkness affects each of us, how it affects and endangers the light of our consciousness and the quality of our soul. We need to recognize how easily it can take away our light and draw us into forgetfulness. And then maybe we can begin to understand what it means for our light and our remembrance of the sacred to be gradually lost, what it means to live in the shadow lands when all that is left are the empty promises of our materialistic dreams.

The myths of humanity have always included the stories of darkness. Sometimes God cleansed the darkness, as in the story of the flood or Sodom and Gomorrah. Sometimes the darkness has a pivotal role to play, as in Christ's betrayal by Judas. Is this present darkness here to destroy the world or precipitate a rebirth of divine love? What are the myths of this present time?

The only certainty I can see is that the darkness *is* telling a story, and that we are a part of this story as is the destiny of our planet. Never has the fate of humanity and the ecological fabric of the whole world been tied together so closely. Never has our forgetfulness of life's sacred nature combined with our greed had such a destructive affect. Our soul and the soul of the world are not separate, and if darkness covers our soul it affects the world soul, the *anima mundi*. We do not know what this could mean, but we can try to listen and be attentive. We can look for the signs within our self and in the world around us. We can relearn the ancient art of witnessing so that we can come to know this story, just as our ancestors watched the stars and the storms, their dreams and the patterns in nature, to know the story that they were living.

What matters with this story is how it is being told in our own individual life. How does this darkness affect us? Does it cover us in forgetfulness, draw us into desires, into its web of materialism, or does it make us feel the sadness of a world that has forgotten its divine nature?

If this darkness does have a mythic or archetypal dimension, as I sense it does, it is important to become aware of the feelings it evokes. Carl Jung stressed that when working with the archetypes[2] it is vital to know the feeling quality of the experience as this adds a human element to the otherwise impersonal world of the archetypes. It stops us from being swallowed into its cold shadows. We do not need to descend into this darkness or try to battle it. This would be dangerous because it is more powerful than any individual. But we need to recognize how this darkness feels within our own life and soul, as well as the life and soul of the world around us.

In my own experience I have watched this darkness grow over the years, seen how it has covered over or devoured a light that belongs to our consciousness of the Divine, of the sacred within ourselves and the world. This has evoked a deep sadness within my own self and soul, knowing how the loss of the light of an individual affects the world soul, the *anima mundi*. Watching this light going out has been like watching a tragedy that few seem to notice, and yet affects us all, individually and as a whole. Without this light to guide us it is much more difficult to find our way, so much easier to get lost. And without this light there can be no real transformation, no possibility of a shift in consciousness, no emergence into any new age.

It is not easy to come to know this darkness. This is a time when it is easier to forget than to remember, to cover our eyes from seeing what is really happening in the inner and outer worlds. But we need to know how the darkness is affecting our planet and ourselves. We need to bring the light of our consciousness and the warmth of feeling into the darkness we are living. Because there is a great danger that once this light has gone—has been completely lost, devoured by the darkness—there will be no remembrance left to tell the true story, no awareness of what is happening. We need to know the story of this darkness before it is too late.

What would happen if life came fully alive, would we know how to respond? Can we regain the ancient, more fluid language of symbols to help us learn to live with real change?

WHAT HAPPENS WHEN *the* ICE MELTS: THE RIVER *of* LIFE *and the* NEED *for a* SYMBOLIC CONSCIOUSNESS

(2010)

A FROZEN SURFACE

While in deep meditation I am drawn into awareness. Rather than dissolving deeper into the emptiness of inner silence, I am asked to listen for a sound, the specific sound of ice cracking. But I can hear nothing, no sound of ice cracking. Then I am shown the image of a river which has been frozen so deeply that it is like solid ground, and it has been frozen for so long that it has been forgotten that it is a river of water. On the banks of the river there is a village or town, and I am left with the thought of what would happen to this town if the ice melted. Would the river rise and flood the houses?

There is no sound of ice cracking, no sign of the river beginning to melt. But the very question, the very suggestion of a frozen river and cracking ice, brings into consciousness a predicament that belongs to life as we know it. This river is the river of life, which has been frozen for so long that we think of life as something solid rather than fluid, which is its natural state. We have forgotten the normal properties of water, how it flows, how it moves and carries us along with it. We are so divorced from any natural understanding that we have constructed our lives, our whole civilization, on a misconception, unaware of the danger that the ice could melt and the river begin to flow again.

Since the seventeenth century when Newtonian physics gave us the basic laws of the physical world, we have developed a scientific understanding and a mechanical view of life as something that can be defined, quantified.[1] Rationalism was elevated over symbolism and we created an approach to life based upon logic and deduction. Recently, the development of computers has generated ever more sophisticated models upon which to base our understanding of our environment and how to plan for the future. They offer the illusion of security, the notion that we can predict what might happen—although the recent suggestions that we may be reaching a "tipping point" of irreversible ecological collapse are beginning to suggest otherwise. In complete contrast stands the ancient Chinese wisdom of the Tao, which taught how to be a part of the flow of life rather than how to protect oneself against unexpected changes. From the same culture come the teachings of the *I Ching*, which explore the very dynamics of change, the understanding that change is a fundamental part of life. This primordial wisdom images life as a constantly changing interrelationship of possibilities, much closer to the image given by particle physics with its suggestion that even the forms of the physical world are just a probability rather than a definable fact.

But today we are living upon a river that has been frozen for centuries, whose ice is solid and deep. There is no natural flow, or even a memory of a time when it was so different. As long as the ice remains frozen we can remain with our image of life as something we can define and plan for, believing we can protect our self by buying insurance against anything unexpected. We feel secure with what appears permanent.[2] The cities we inhabit are designed to last like the concrete of which they are constructed, rather than to constantly adapt and change. What is fluid we are not prepared for. We are taught how to live with facts rather than probabilities.

Part of the problem comes from the very way we think and are educated to think, and in particular how we have banned symbolic consciousness. We are taught to think in an analytic, linear manner, using words to explain our self. Symbolic consciousness is holistic rather than analytic, and rather than thinking in words it thinks in symbols and images. It was prevalent in our Western consciousness as recently as the medieval period, as expressed in the sacred geometry and iconography of the Gothic cathedrals.[3] Symbols connect us to the interior world of the soul, and symbolic consciousness enables us to realize the sacred meaning that underlies our physical existence. In symbolic consciousness everything is part of a pattern of interrelationship connecting the visible and invisible worlds; and, as anyone knows who has worked with dreams and their symbols, this is a very fluid, amorphous language, in which images change and evolve, giving us possibilities of meaning rather than definable facts.

Symbolic consciousness was central to human consciousness for thousands of years.[4] We lived in both the inner and outer worlds without any contradiction. Shamanic wisdom carried the understanding of how these worlds interrelate, how they reflect and flow into one another, and the destiny of a tribe could depend upon a dream. Symbolic consciousness presents a worldview so different from our present model that it is difficult for contemporary consciousness to grasp how much it was once a part of everyday life. We do not realize the limitations of our rational consciousness, or how we have become caught in its constrictions without even knowing that we have lost part of our natural awareness. We believe the facts with which we are presented, without fully recognizing that they are only a probability, and that nothing is fixed or definite. We have become strangers to the symbolic world, and we have lost the fluidity of consciousness that belongs to this more primal awareness.

Symbolic consciousness is a part of our natural relationship with the soul and the sacred that is present in all of creation. It connects us to a world full of meaning and wonder. Rational consciousness instead imposes its vision of reality, instructing us with the laws and scientific principles that now define our life. Initially rational consciousness was seen as an "enlightened view" that could free humanity from the "darkness" and fears that had imprisoned us in superstition.[5] But it has been imposed so successfully that we are no longer aware of what it is excluding, of the sacrifice of the symbolic and our connection to the sacred. It has cut our consciousness off at the roots so that we no longer have any natural connection to the mystery and joy of life.

The question that then needs to be asked is whether the icebound river is just part of the flow of the ages: a time of winter that has lasted for centuries. Or has the development of rational, analytic consciousness itself produced this frozen landscape? Particle physics has proved what Buddhist teachings have long known, that mind and matter are not separate but influence each other. If the reality we inhabit is created by our consciousness, then we could have frozen the flow of life with our vision of a solid, definable world. We would then be like the ice queen who has turned the world to winter. And now we inhabit this desolate landscape where joy and symbolic meaning lie hidden beneath the ice.

Symbolic consciousness allows for a deeper understanding of life, with all its patterns of interrelationship, than does a purely rational approach. Symbolic consciousness has given me an image of the river of life as frozen, and asked me to listen to the sound of the ice cracking. But there was no sound, nothing. All I was left with was the thought of what would happen to the town on the banks of the river if the ice were to melt. In this symbolic picture there is no solution. It just gives an image of a different perspective on life, suggesting that something so fundamental to our existence as the river of life is no longer flowing—and also that we do not remember that the real nature of the ground upon which we live is not solid. Maybe we have based our whole civilization upon mistaking a temporary state for something permanent. And we have not begun to question this self-imposed belief.

MOVEMENT IN THE DEPTHS

But the river itself remains alive. It may be frozen, waiting for a thaw, but it still carries the energy of life. We exist so much on the surface of life that we have little understanding of its depths and the currents that run there. Our lack of a symbolic consciousness has not only isolated us from the flow of life, but also cut us off from its depths—the primal, archetypal depths of life that have always communicated to us through images rather than words. And even though the ice is not yet even cracking, there are changes taking place deep under the surface. The energy of life is beginning to flow in a new way, to follow different patterns. Carl Jung described these archetypal patterns as the riverbeds through which the waters of life flow. And the archetypes are shifting, some awakening from a long slumber. They are beginning to move in new ways: a new energy grid is constellating.

Our knowledge of history is so recent and so censored that we cannot imagine what might happen if these energy patterns of the deep change. We live so much on the surface that we have lost any knowledge of the depths of life, of the energies that underlie our existence and how they affect us. We also live with the illusion that we determine our own future, create our own destiny. We may feel that something fundamental in our lives is shifting, experience an unexplained insecurity. We may look for predictions and even prophecies to comfort us. But the movement in the depths of the river of life is real, even if it is still hidden under the ice. And when the ice breaks the river will carry us along however much we resist. We are a part of life, even if we have tried to separate our self from its primal energies, built our cities and towns to protect our self from the forces of nature.

For those of us who are awake to the symbolic world, our work is simple: to listen and watch with a consciousness that can be attuned to the depths. There are signs all around us, and in our dreams there are messages of meaning. We do not yet need to "interpret" what they say, because what is more important is that we listen to this symbolic language, be receptive to its images, and in so doing attune our self to the energy that underlies life. The symbols themselves will reconnect us, because this is a part of their function. And through this reconnection we will come to know what is happening in the core of existence, in the sacred depths of our being. Life itself will tell us what we need to know, just as it has communed with human consciousness for millennia. We just need to be attentive and listen: then we will feel how the currents are changing and what this will mean to our surface lives.

Only if we reconnect to the sacred core of our being will we have any understanding of what is happening. Because it is here, where the divine energy comes into manifestation, that the real changes are taking place. All energy, all real change, comes from within, from the Divine of which outer life is a manifestation. It is partly our collective separation from this center that has caused the river to freeze. It is our denial of the Divine that has isolated us on the surface of life. The changes taking place are a reawakening to what is real, to the sacred that is within all of creation. But in order to read the signs of this reawakening we need to relearn the language of the sacred. This is of primary importance. Only then can we play our part and welcome the waters as they start to flow.

Our hesitation will come from our holding on to the image of life as frozen, as something solid. Sadly many of the skills we have learned and technologies we have developed belong to this image of life, and will be as useless as a car without gas. We will have to relearn many skills, change many of our attitudes. We will have to relinquish many of our patterns of control, our images of power. We will also have to learn again how to live with the Divine not as some transcendent being but as a real presence and energy that is central to our existence. And we will have to learn what it means when the waters of life start to move.

Many things that we thought valuable may be lost in this flood. Maybe even the towns upon the banks of the river will have to be sacrificed to the water. They were built without any understanding of the real, volatile nature of the river. We cannot afford to spend too much energy protecting our property and possessions, because then we will miss the opportunity of movement, of where the water can take us. We will get caught in a toxic backwater slowly dying. Life is about change and learning how to be with the energy of change. It is not about protecting our self from the future. The changes will bring possibilities we cannot imagine, and also bring their own dangers. How we adapt to the awakening flow of life will determine the future of humanity.

But at present the work is to learn to listen to the signs, even if as yet there is no sound of the ice cracking. We need to regain our symbolic understanding, because it is in these images that the book of life is being written around and within us—"We will show them Our signs on the horizons and in themselves."[6] This is the first step to take: to reconnect with this ancient language of the sacred, where the Divine and human meet. Only then can we begin to understand what is happening.

For so long we have separated ourselves from the world of light. How can we remove this deep division and claim the light we so desperately need at this time? First we need to understand the source of this separation, how humanity chose to create a barrier between the worlds.

THE WALL
(2010)

A WORLD OF LIGHT

In deep meditation I come to a wall. I know this wall. I have
seen it many times before in meditation and waking visions.
It is a high brick wall. I know what is on the other side of the
wall: a world of light. But there is no way through; there is
no doorway, no ladder, no break in the wall. When I come to
the wall I walk along it, and then I have to turn away, back
to the narrow streets of this world. And yet I know what is
on the other side. Sometimes I have made every effort and,
clambering to the top, looked over the wall. Or I have just
felt what is there—endless expanses of light, and the beings
of light who live there. And yet always I have to come back,
back into this world, so constricted and full of shadows:
the half-light of our existence.

In the summer of 2008, I spent three weeks on the other side, in that world of light. It was a crazy time. I was very ill and hardly slept. When I went to bed and closed my eyes I was in the world of light. There was no need to sleep, no possibility of sleep. There was so much light; there were experiences in light. Light upon light. Sometimes during the day, too, I was fully awake in this world of light. I could see our world from the other side, see its loves and hopes and dreams, its worldly power structures and places of prayer. I could see the spiritual essence of every tree and flower and the patterns of darkness in which people are so caught. And I saw the beings of light that are waiting for us, that want to help us, and I saw how we have forgotten them. I saw this sticky substance of forgetfulness that covers us and drains away any remembrance we may have. And I saw how other beings of darkness that belong to this world also drain our light, keep us caught, cover us in greed and desire, hatred and anger. And I saw that this is how it is.

But I could not live forever in this world of light, even though I longed to. There was too much light. It burned into my consciousness. It did not allow me to sleep. I was exhausted. I needed to be able to live in this world, however dense and distorted. And so in order to survive, in order to live, I turned away from the world of light. I closed my consciousness to it and focused on the physical world, on letting my body heal. For weeks I hardly prayed or meditated. I worked on my house, focusing on the walls, doors, and ceilings I was painting, rather than on endless horizons of light. I came back into this world, battered and bruised, sometimes full of resentment at having to leave behind the light, feeling angry, deserted, and betrayed at having to return. How could I be given a taste of the beyond and then be pushed back into the darkness and limitation of this world with all its distortions and misunderstandings, all of the stuff we have been conditioned to call life? Yes, there is beauty here, but there is so much darkness. On the other side there is not this darkness, or this density; there is never this forgetfulness. We are beings of light. How can we forget?

Once before, when I was twenty-three, during a summer of intense inner experiences, I was taken to the other side and given the choice to live or die. I remember this experience so vividly: being taken out of my body, high up into a place of freedom and light. I was told very clearly, "You are free now. You can go." And just as clearly I remember my reply: "I am a Sufi. I am here to be of service." And so I returned. You never forget the consciousness of the other side. It haunts you as both a promise and a poison. Sometimes it makes you long for death, to return to the light and freedom that you know are waiting. But of my free will I had made a pledge, a promise; and so I returned and the real spiritual training began.

And now, over thirty years later, I was taken back to the other side, and not just in a moment in and out of time. First

I had an inner experience that was like tasting death, with all the suffering that so often accompanies death, and then I was in the beyond. For three weeks I was fully conscious on the other side. And what is beyond is so pure, so endless and unencumbered. There are oceans of love and light. And then I had to come back. I knew I had to come back. My body and mind could not live any longer in that world of light, and it was not yet time for me to fully die. I knew this, even as I longed to be released, even as I pleaded and complained, was angry and resentful. I had to come back. My body and mind had to be healed from being exposed to such an intensity of light. And afterwards, over the weeks and months that followed, this is what slowly happened. I did not want to be here anymore. And yet I was here, and the world still looked the same as I remembered it. There was no profound insight, no illumination. I had been on the other side and now I was back, battered, bruised, and complaining. But I was back.

And I felt tired. Even when the body was healed I felt tired. Was this tiredness just the aftereffect of this experience, or was it something else? I began to realize that it was a deep tiredness of the soul rather than of the body. It is as if the very fabric of my innermost being was exhausted. And I wondered, how can the soul be tired, when the soul belongs in a world of light and love, when the soul is the part of our self that is with God?

And then I saw the brick wall. I knew what was on the other side. This time I had no need to try to climb the wall, to look over. I knew that landscape of light, and how different it is from what we call life or existence. And I was left on this side of the wall, in these narrow streets, where even the flowers that grow beside the streets have forgotten the world of light. And I wondered, has it always been like this? Is this just the wall that separates what we call life from what we call death? I know in meditation one can leave behind the body and the mind and go into the light, but always one has to come back. Is the only way to fully live in the light to leave the physical world behind and die? Most people only access this world of light after they die, or in near-death experiences. Is this wall the barrier that has been placed between the worlds, like the river Styx of the ancients?

The Sufis describe how we need a separation between the worlds, "seventy veils of light and darkness," or "the glories of His Face would burn away everything." As I know from my own experience, the light of the Divine is too dazzling for us to perceive directly; its energy is too strong. This is one of the reasons why spiritual life is a slow process, a gradual lifting of the veils as one develops spiritual strength, becomes more and more able to bear the light. But these veils filter the light. They are not a wall that cuts us off from it.

Now, for the first time since I had seen this brick wall, I began to wonder. Why was I shown it like this, always a brick wall? It is not a river of forgetfulness, a veil of light, or a rainbow bridge. It is made of bricks, and bricks belong to this world. And then suddenly it dawned on me. This wall was made brick by brick by human beings. It is not a natural separation between the worlds. It had been purposefully built by people, by their ideologies, laws, and power structures. Humanity had purposely created a wall of separation between this physical world and the world of light; and it was built so long ago and has been so effective that we all accept it. Now we live in the shadow of this wall without even noticing it. Nor do we realize that we have been denied our heritage of light. We have been conditioned to accept the world of shadows and half-truths we call life without even realizing that we are cut off from the world of light. This has become our heritage. We have successfully stranded ourselves from the Divine.

TIREDNESS OF THE SOUL

And suddenly I understood this deep tiredness of the soul, the tiredness of living in this world cut off from the light, divorced from what is real. I realized how deeply this light sustains us, nourishing our real spiritual being. In the depths of our being we are made of light, we are beings of light.

Light needs light. We need this light to nourish us; otherwise our soul is left hungry, even starving. Without light we can become spiritually exhausted, listless, depressed. Our divine nature needs this inner light as much as our physical body needs food and sunlight.

As long as we are pursuing the desires of the ego, this world has an energy that sustains us. It is full of desires to attract us, and the instinctual energies of life draw us into its journey of self-discovery. We also have a culture that supports us with endless attractions and addictions, full of promises of fulfillment that lure us into believing they will actually be fulfilled. But there comes the time when the individual tires of this endless pursuit of self-gratification, when the soul prompts us to make the journey of return, to discover the light that is within us. And this is the real and demanding journey that is called spiritual life, during which gradually we access the light of the soul and are nourished by it. Life seems to transform, and we are nourished by the meaning of the soul rather than the pleasure or pain of desires. But if one looks closely one will find that something is missing, both within our self and within the world around us. And what is missing is a certain light that should also belong to life, a certain note in the song of creation that should be present. And this light, this note, is the deep knowing that everything belongs to God and is an expression of the Divine. And despite the beauty of the world, despite its horror, we have lost this knowing: its note is hardly present, its light has gone dim. And this is the unspoken tragedy of our world.

And when you begin to wonder how you can live in a world where this note is not heard, even the promise of inner fulfillment falls away. What does anything really matter if this primal truth is denied? I realized that this truth has been denied to such a degree that we no longer even ask the question. We ask many questions about the state of our world, about its poverty and ecological devastation. But our consciousness has been so effectively censored that we do not ask the most important question—what happened to the Divine? Where in the world is the light that belongs to God? If we are really concerned about the state of the world, this question is even more pressing, because nothing can really be done without this light, this energy, this power that comes from the Source. But we have been made to forget that this light even exists. The wall that separates us from divine power has been present for so long that we no longer ask what is on the other side. Nor do we realize that it is just a wall. We have been betrayed by the power structures and ideologies of this world more than we can imagine. And we are a part of this betrayal. Our forgetfulness of the light reinforces the wall, strengthens its bricks and the mortar that cements them in place.

Deep in the soul there is this exhaustion, as if the soul of the world itself can no longer bear this desert of separation, can no longer sustain itself without the light. For how much longer can we continue? Can we continue forever without this light?

THE CREATION OF THE WALL

I began to wonder how the wall was created. There was a time long ago when humanity lived in a world of light. Maybe this was what we call the "Golden Age." It may be likened to the mythical Garden of Eden in the Bible when Adam and Eve walked naked in the presence of God, before they "hid themselves from the presence of the Lord God" (Genesis 3:8) and were then cast out. At this time there was no wall between the worlds: we were not separate from God. We lived in divine presence and had no consciousness of any other way of life. But then, in contrast to indigenous cultures who live in harmony with the natural world and its spiritual dimension, in our Judeo-Christian heritage humanity began to claim the power of the ego and became separate from God, experienced the "Fall."[1] In the Bible this is imaged as eating of the forbidden fruit. Because humanity had been given the gift of free will this transgression was allowed to take place. We were allowed to go against the law of God.[2]

It is important to realize that this first act of transgression was a *conscious* choice. The Bible may blame Eve for seducing Adam, but behind this patriarchal dynamic is the awareness of a choice to go against the law of God, to eat of the forbidden tree. This is why we were cast out: we *chose* to deny the divine law, and this choice has continued throughout the millennia. Humanity liked the feeling of its own power and autonomy and created a separate world governed by its ego.[3] Turning our attention away from the Divine and focusing more on our own power, we began to build a wall between the worlds. Gradually we created a world in which we experienced the Divine as no longer present. It is our denial of the Divine that has crafted and created this wall of separation, that has banned us from our spiritual heritage.

At different times saints and sages have come to remind us of our divine nature. Christianity was born through an awakening of divine love with its message of sacrifice, forgiveness, and mercy. Through Christ's life, crucifixion, and teachings, the gates of grace were opened, and light and love flowed into the world. In the early years of Christianity, through the devotions of small circles of believers there was a continued outpouring of love. But only too soon the mechanisms of worldly power began to block this flow. Under the guise of unifying Christian beliefs, those teachings that gave the individual direct access to the Divine, for example those in the Gnostic Gospels, were banned, and the practitioners were persecuted as heretics. Only through the priests and the hierarchical structure of the Church could the individual have access to God. Gradually, but also systematically, as the Church became a temporal power, religious structures were created that strove to keep God in heaven so that the church hierarchy could keep its power on Earth. Finally, in the brutal slaughter of the Crusades and the tortures of the Inquisition, we see a Church that has chosen to abandon love and forgiveness for the fruits of worldly power. What is less understood is how this religious ideology cemented the separation between the worlds. God could only be reached after death; heaven could not exist in this sinful world.

In the West since the "Age of Enlightenment," rationalism and the pursuit of science continued to reinforce the wall of separation as the world came to be seen as a mechanical place devoid of any sacred nature. The sacred groves had long been cut down by the patriarchy, Christianity had worked to eradicate any pagan beliefs[4], and science now gave us an unfeeling, barren world that it could conquer with technology. The wall between the worlds became so much part of humanity's consciousness that in the West we no longer knew there was a wall. The fact that the world was starving from a lack of the sacred did not even enter into our collective awareness. Finally, in the last century, communism and capitalism became the twin demons of the world, each celebrating an existence defined only by what we can see and touch. And when consumerism triumphed and the glitter of its toys captured our complete attention, no one seemed to notice that the Divine was not present. We had lived with the wall for so long there was nothing in our collective memory to remind us of what we had abandoned, of what is so close on the other side of its bricks.

Of course there have always been individuals who, either within or outside the structures of religion, have consciously made the journey to the other side. Humanity has always had access to spiritual techniques to directly access the light. The discipline of meditation, for example, is a simple practice of turning within and accessing the light through one's higher spiritual centers. If one accesses a higher consciousness within oneself there is no longer a wall: one is present in the dimension of light upon light. This is the pure consciousness realized through mystical practice, such as Buddhist meditation, or the light of the heart of the Sufi path. These are beautiful and powerful practices, through which one can *transcend* the limitations of the physical world while still present in this world. But the focus of these practices has almost always been of an *ascent,* of going beyond the physical world with its sufferings and problems. Their effect is usually to draw one away from the outer world. It is easy then to become detached, no longer interested in the demands of everyday life, which through the practices have become illusory. These practices do not dismantle the wall: instead they offer a way to go beyond it, even to give one access to a consciousness where the wall does not exist, where there is no separation between the worlds. Mystics and spiritual travelers who have had these experiences may remind a few people that this world of light exists and that there is a way beyond the wall into the light, but the wall remains as solid as before. And the world in which we live today is left starving for light.

A LIGHT TO SEE WHAT IS REAL

Although no one can now remember it, the light does more than just nourish us. This light enables us to see what is real. When it used to shine in the world, it revealed the true nature of things, their real purpose and meaning. In this light we could each live according to our true nature and recognize the true nature of others and of the world around us. We could see the world as it really is, the divine creation of which we are a part. The world thus seen, in its real nature, is quite different from the world created by our desires and projections, by the endless patterns of our mind and the recycling of our memories that we call existence. Anyone who has for an instant awakened, had a glimpse of what the Zen masters call *satori*, will know this simple experience of truth, when the butterfly is glimpsed as a butterfly, the plum tasted in its sweetness. It is a reality without comparison or contradiction that communicates its true nature to us directly rather than being interpreted through our mind or psyche. In such moments we are really alive, awake rather than dreaming.

Long ago, at the dawn of consciousness, humanity was given the ability to recognize and relate to the true nature of everything. This ability belongs to the naming of things, because everything that is a divine creation has a name that embodies its true nature. In the Qur'an (2:31) it is written that "He taught Adam the names/Of all things," meaning that Adam was taught the inner nature and qualities of things.[5] And so the first man had knowledge of the names of creation, which belong to the divine "secrets of heaven and earth" (Qur'an 2:33).[6] This knowledge of the inner nature and true purpose of the created world is part of our divine heritage, our inner Adam. It enables us to participate in life *as it really is*, as a divine revelation: the world created by God rather than the world created by humanity. But this knowing can only be accessed through the light of our own true self. Without this divine light we cannot read the book of life. We do not know what is real, or what is the true purpose of our life. Today we have long forgotten the true names of creation. We remain caught in the surface patterns of illusion, in our projections and fantasies. This is our present collective predicament.

There are always some individuals who are able to awake to the light of their true self, to see what is real and how to live the meaning of their soul. But for our Western collective this is a distant myth. Many of us wander through our lives often becoming lost in a world without true meaning, unable to find our way. There are always signs that point towards what is real, but we cannot recognize or read them. It is only too easy then for the power dynamics of the world to delude us, to trap and enslave us. Without any real knowing, how can we find our way out of this maze in which we serve the lords of this world rather than our true Lord? Once we have access to real light we can see how we have been deluded, sold our soul for a few pieces of silver. But without the light we know nothing of the world around us: we only see the images that have been made to glitter. We are caught in the illusions that are spun around us.

This is partly why the powers of this world want to deny us access to the light, want us to remain in the shadows. Without the light we are easier to mislead and control, and can be sold worthless trinkets. We know nothing and see nothing, and easily believe what we are told.

DESTROYING THE WALL

But now the deeper question remains, do we have to remain stranded in the shadows when the light is so near? Is it our collective destiny to be imprisoned by the wall of separation created by our ancestors and reinforced by our own forgetfulness? Or can we reclaim this dimension of light, the light that will enable us to see the world around us, a world of beauty and wonder in which the Divine is present, where joy has returned and we no longer pollute our environment with our unsustainable desires?

How can the wall be destroyed? When I look at the wall I see no sign of any attempt to break through it. There is no indication of any real rebellion. There are no armies of saboteurs attacking the wall, or even ladders placed against it. Its bricks are smooth and polished. The wall appears untouched. We seem to have accepted the wall without question. It is high enough that we cannot see over to the other side. It has been there for so long and become so familiar we do not even notice it. Who is there to question it? Do we have the will or the power to destroy this barrier to the light? We are so seduced and drugged by the playthings of this world that we do not question what we have been denied.

And yet our own soul and the very soul of the world are crying out for the light. Without it we cannot heal or redeem our dying world; the sacred and the remembrance of the names cannot return. Only with the light can real meaning reappear.

Leonard Cohen once wrote, "There is a crack in everything. That's how the light gets in."[7] Perhaps we need just to make a crack in the wall through which the light can then begin to stream. Many times cracks have appeared in our collective defenses, as for example in the "flower power" movement of the 1970s with its vision of peace and love. But these harbingers of light do not appear to last. They self-destruct, destroyed by drugs, for example, or are swallowed back into the collective, sold out to materialistic values. Sadly, much of the new-age spirituality that brought the light and the practices of spiritual traditions from the East soon became corrupted and self-serving, using the energy of the light for money and ego desires. The forces of darkness know our weaknesses only too well, and the cracks are quickly mortared over before enough light can come through to make a real difference. Maybe a crack in the wall is not enough.

If the Earth itself needs this light, can the Earth rebel?[8] Can the primal powers that are present within creation awaken and destroy what humanity has created? This is a possibility, though it would be very destructive. We may have denied and forgotten the dragon-powers of creation, the archetypal energies that underlie all of life, but this does not mean that they no longer exist. Mostly they are sleeping, but this does not mean that they cannot awaken, and like the gods of old react with power and violence. The wrath of the gods is not just a myth.

What would it mean if the energies of the Earth were to reclaim their connection to the light? In order to destroy the wall of separation, how much of humanity and its images of civilization would be devastated in the process? The light could return, but who would be here to welcome it? Who would be here to remember the names of creation and the sacred ways of working with light? Today's speeding up of our ecological imbalance could be signs of a world grown tired of waiting for humanity to take a step. We may now recognize that there is a global ecological crisis, but who, apart from a few shamans, knows how this relates to the powers within creation? Our scientific models cannot begin to understand what belongs to the depths or how it would interact with our surface life. Mostly we remain arrogant in our ignorance; and yet there is also a deep anxiety now present within the collective that tells another story, as if the collective itself knows that there is a storm coming, one that its politicians cannot prepare for.

Then there is another possibility, more wonderful and awe-full than anything we can imagine. What if the wall were destroyed from the other side, by the energy of light itself, and by the beings that are in service to the light? Within the world of light there is more power than anything in this world. This is one of the reasons the light has been veiled from us—it easily overwhelms us with its power and glory. Anyone who has encountered the light knows this: how a few moments of light are all that one can bear, how just one glimpse can change one's life forever. And this is just a particle of the world of light. If the powers of light were to return to this world, it would be like stepping from a darkened room into brilliant sunshine—blinding and beautiful.

Having lived in this world of light I know its beauty and power, its simplicity and love. This is a dimension of clarity, without the distortions and confusions of our world. And it is governed directly by divine will and divine law, without the intercession of human will with all of its mistakes and power dynamics. The Sufis call this dimension "the world of divine command" (*'âlam al-amr*)—in contrast to "the world of creation" (*'âlam al-khalaq*), which we experience through the senses and the veils of the ego. In the world of divine command everything bows down before God. It is the domain of angels and other beings of light who only know to bow down before God, and can only enact God's power and divine will.

What would happen if the power of light and the enactment of divine will were to return directly to our world with all of its dramas of worldly power and fantasies of self-empowerment? How would we respond? Would we know to bow down before That Glory, or would we rebel and fight, try to hold on to our images of ego-power, and become caught in a battle of light and darkness? Would a deep inner knowing of the reality of divine will surface into our collective consciousness, or would we see this power as another aggressor whom we need to fight to keep our independence? Do we even know how to bow before God and the messengers of light, or are our images of personal freedom too important?

It is so long since the light was here that we have almost no memories even in our ancestral consciousness of how to live in the light. We have learned how to live in the shadow lands of our culture, how to manipulate and deceive, how to protect our self and our possessions. But in the light there can be no manipulation or deception: there is too much light. We will have to learn once again how to be honest and truthful, how to be sincere and open. And how to take real responsibility. This is the only way to live in the light.

If it is the will of God the wall could be destroyed by light. It could dissolve in an instant. But how will humanity bear the light when it knows only shadows? How could our consciousness cope with a world of light? I know from my own experience how difficult it is to contain this light while living in the physical world. There need to be veils between the worlds, but veils that filter the light, not walls that stop the light.

If the wall were to be removed, would it be easier for it to be slowly dissolved, so that the light could gradually come into this world? Or would the forces of darkness and worldly power just mobilize themselves to repair the wall and stop the flow of light? These are questions we cannot answer. But we can recognize that there is a world of light that does not belong just to some unobtainable heaven or elevated spiritual state. It is here, just beyond a wall that we have built with our own ideologies and patterns of control. And we know in our depths that humanity and the world itself cannot survive much longer without the light that comes directly from the Source. Everything else has become too polluted and corrupted. The heart of the world is bleeding, and the soul of humanity is crying out. We need this light in order to see our true nature and the true nature of life. And life needs this light in order to heal and transform, so that together we can make the next step in our evolution.

As humanity made its collective choice of economic progress over any real concern for the Earth, I began to witness in deep meditation what I did not want to bring into ordinary consciousness—it was too painful and desperately sad. I saw the light going out, the possibility for a real global shift being lost.

WITNESSING *the* END *of an* ERA
(2011)

In March 2011, the science journal *Nature* announced what many people already knew, that there are clear indications that the world's Sixth Mass Extinction is already underway. The last mass extinction came some 65 million years ago when a comet or asteroid slammed into the Yucatan peninsula in modern-day Mexico, causing firestorms whose dust cooled the planet, and an estimated 76 percent of species were killed, including the dinosaurs. The four previous mass extinctions of species were due to gradual global warming and cooling, and happened on a scale of hundreds of thousands to millions of years. What is particular about our present mass extinction is that it has happened so quickly over a few centuries, and most significantly, it is man-made. "The modern global mass extinction is a largely unaddressed hazard of climate change and human activities."[1]

We are slowly, and in some cases reluctantly, waking up to the global ecological disaster of toxic pollution, climate change, and extinction of species. And the fact that these changes may be irreversible and possibly catastrophic is becoming known. The Buddhist ecologist Joanna Macy suggests that we may be coming to the end of the Cenozoic Era, the 65 million years since the last Mass Extinction:

> Continuing on our "business-as-usual" trajectory will acidify the oceans and trigger runaway global heating, epic mass extinction and a completely new cycle of geological time. A few climate scientists consider we may have already entered into runaway climate change.[2]

Humanity is now faced with this unprecedented, self-created global disaster, and as the recent climate change conferences in Copenhagen (2009) and Rio (2012) illustrated, we seem unwilling to face the facts, putting short-term economic gain above the reality of what is actually taking place and its long-term consequences.

However there is another dimension to this global predicament that we should not ignore. Many spiritual teachings tell us that what happens in the outer world first takes place in the inner. We know this in our individual practice, as expressed in the simple mantra "all change happens from within." Indigenous peoples have long understood how this happens on a macrocosmic level, and their shamans often worked to keep the inner worlds aligned so that the harmony of life could continue. Martin Prechtel describes how their work was to heal any imbalance that we might create in the web of life:

> Shamans are sometimes considered healers or doctors, but really they are people who deal with the tears and holes we create in the net of life, the damage that we all cause in our search for survival.[3]

If the outer, material world is a reflection of the inner world, we must both "face the facts" and take real responsibility for our outer situation while at the same time acknowledge the damage that we are causing in the inner world—in our own souls and the soul of the world.

While an indigenous culture and its shamans would look to the inner in order to understand the outer, this is not a part of our Western spiritual heritage. And while Eastern spiritual traditions have helped us to understand that personal transformation depends upon inner change, the larger, macrocosmic dimensions of these teachings have been mostly overlooked. How do we approach this task for which we, as a culture, have so little understanding or training?

THE CONSCIOUSNESS OF SEPARATION

It should be understood that the inner worlds are influenced by the consciousness of humanity, which is why spiritual teachings throughout the ages have stressed the importance of our attitudes and the values by which we live.[4] Thus, while our physical acts affect the outer world and have produced our ecological crisis, it is our consciousness—our attitudes and beliefs—that directly affect the inner, spiritual dimension of life on Earth.

One of the most damaging attitudes of recent centuries is the attitude of separation. Since the Age of Enlightenment our Western culture has held the belief that humanity is fundamentally separate and distinct from the physical world around us—we do not think of our self as part of the living being of the Earth. Modern physics tells us that the universe is a unity: that it is undivided, and that beneath the surface every object and event in the universe is completely woven up with every other object and event. Similarly ecology has revealed that we belong to an interconnected, interrelating environment. However our consciousness and attitude towards the world is still conditioned by a Newtonian paradigm of separation, an attitude that seeks to dominate and control nature through science and technology.

This mechanistic worldview has been combined with a religious belief that locates God above or beyond the natural world, and in the case of Christianity, has sought to erase "pagan" spirituality, which recognized the sacredness of nature and its cycles. These fundamental attitudes have allowed us to abuse and pollute a world of unfeeling matter.

For many years I have witnessed how our collective attitudes, our disregard and forgetfulness of the sacred, our focus solely on material well-being, have been very destructive. I have watched how humanity's sense of the Earth as "other" has allowed us to treat the Earth as a means to an end, as a resource to be exploited. And I have seen the impacts of these attitudes on the inner worlds. We have polluted not just our rivers but also the sacred waters of the inner world. Places of power and beauty have been spoiled, their wisdom and magic covered over or lost.

I have also seen the inner result of our neglect of the world soul, the *anima mundi*, which the ancients understood as the spiritual ordering principle in creation.[5] A desolation and deep sadness are now present in the inner world, as what is so essential and precious within life has been abandoned.

During the last year I have become aware of an even more disastrous change. A light in the inner world that gave meaning and spiritual sustenance to our souls and to the whole world has been going out. And it is now extinguished.

Something that for millennia was central to the inner life has gone, lost through our greed and arrogance, our ego-centered power dynamics and forgetfulness of the sacred. We are not entering just an external era of extinction, but an inner dark age. And what is more dangerous is that we do not appear to know it is happening, even though this inner light is fundamental to the well-being of our individual soul and the soul of the world.

LIGHT AND MEANING—THE INDIVIDUAL

Our Western collective attitudes and beliefs about the Earth have desecrated the Earth and its spiritual nature. As we have become more and more seduced by the soulless values of consumerism, this deep forgetfulness of the sacred nature of creation has accelerated. It has become like a dark cloud that has covered over and finally extinguished a light within the inner world.

This is a global phenomenon. As the light of an entire era has faded, our global culture has lost something essential.

Why have we let this happen? Partly it is because we simply do not understand the relationship between the inner and outer worlds—how an inner light gives outer life meaning, how we are sustained and nourished by energy from the inner world that in turn nourishes our outer life. And we don't understand that this relational and mutually dependent dynamic of inner and outer takes place both on the individual and collective levels, how microcosm and macrocosm reflect each other. In a culture that focuses so much on the individual, it can help to look at the relationship of light and darkness that takes place within every human being.

In each of us there is a divine spark that belongs to our soul, which, if nurtured and cultivated, can enable us to develop to our full potential. This light is what gives direction to our life so that we can live the meaning of our soul, the deepest purpose of our incarnation. Most people experience this as the light of their conscience or moral compass that guides them through life. As we follow the deep inner knowing of the soul and its promptings, we are drawn to situations that help engage us deeply, help us learn and develop, and offer opportunities for us to realize our unique potential.

Those drawn to a spiritual path are able to work with this light. Through purification and spiritual practice we are able to increase our light and have more direct access to it. The spark becomes a flame that burns more and more brightly, enabling us to see more easily the path we need to follow back to our real Self.[6]

As we honor and follow this light, the easier it is to live the purpose and destiny of our soul, and thus have a deeper and more meaningful life. It is this light that enables us to transform ourselves and uncover what is real within our heart and soul.

However, selfishness, forgetfulness, and the many corrupting influences of life easily cover over our individual light. As the light fades, we become more easily trapped in life's many illusions, unable to change or evolve.

This journey of a soul into darkness, cut off from his or her light, has been told through myths and stories for ages: for example, in the figures of Faust or Macbeth. One or a series of actions—motivated, for example, by greed rather than generosity—puts the tragic hero on a path of loss and destruction until an opportunity for redemption is offered. We know this trajectory well, how easy it is to slip away from the light into a darkness where we often feel that we have lost our way. Then—perhaps—by some twist of fate or grace, we wake up to our situation and find a second chance to do what is needed and repair what has been damaged. Or tragically, the offer of redemption is not taken, and the light of the soul becomes lost. Sometimes one can recognize this story in an individual, when in their eyes we witness the loss of the light of the soul, sold through desire for power or many of the ways we chose darkness over light. Something essential to our human nature is no longer present.

The ways the light of a human soul is honored and strengthened by moral choice, right action, or spiritual practice is a mirror for how an entire civilization can work with its collective light. So too, the ways that an individual human being can turn from his or her own conscience or inner wisdom, lose their light through selfishness, is a mirror for a collective forgetting.

LIGHT AND MEANING OF AN ERA

What is true for the life of an individual is also true for humanity as a whole. Each era has a particular light that enables humanity to fulfill the purpose of that era. From a spiritual perspective each era of humanity has a spiritual light that gives rise to the evolution of consciousness that belongs to that time.

The real evolution of humanity is the evolution of consciousness,[7] and it is the light of the inner world that both guides and facilitates this evolution. Humanity has a unique quality of consciousness—mythologically imaged as fire stolen from the gods—that is quite distinct from the instinctual consciousness that belongs to the animal kingdom. Throughout human history our consciousness has evolved and changed, and brought with it changes to our outer life and ways of worship. In each era this inner light has particular qualities that enable human consciousness to change and evolve in a specific way. In our present era that is coming to an end, this light has enabled us to develop an understanding of the physical world that has fundamentally changed our human experience.

One way this inner spiritual light directly manifests itself is through human discoveries. For example, the Gothic Age was made possible through an understanding of how stone can create cathedrals of stained-glass and light that were almost unimaginable before. The Gothic Age was an outpouring of inner energy that resulted in both a spiritual

119

resurgence and what is arguably the most beautiful architecture ever created in Europe. Similarly the Renaissance
was a development of consciousness celebrating the human
spirit in an entirely new way through the medium of art
and architecture.

Since the eighteenth-century Age of Enlightenment we
have been given many discoveries, particularly in the fields of
science, that have helped to create our present civilization. We
may think that these discoveries were human achievements,
but from a spiritual perspective they happened through a
moment of "illumination" when a light from the inner world
resulted in an inspiration or understanding. This is clearly
illustrated by the story of Newton "discovering" the law of
gravity in the instant when the apple fell on his head. The
"eureka moment," as scientists call it, indicates a moment of
grace—when thought and effort cease, the veils are lifted,
and truth is suddenly available.

Behind any great human discovery is the inner light that
gives birth to inspiration. In the last era this light carried a
quality, or knowledge, of an understanding of the physical
world and the properties of matter. Through this light human
consciousness has been given an understanding of matter
that it did not have in earlier eras. In previous eras human
consciousness evolved through an understanding of other
realities: for example, through a knowledge of the interior
world of symbols and their magical properties, much of
which knowledge we have now lost.

Sadly, while the earlier discoveries combined spirit and matter—for example, the principles of sacred geometry helped align the spiritual and physical worlds[8]—more recent knowledge appears to have come at the cost of the sacred. Newton may have been fascinated by magic and alchemy, but his "laws" gave birth to a vision of the world as a machine without a soul. It is this soulless world, cut off from an understanding of the sacredness of creation, that our civilization now inhabits.

As humanity developed a consciousness that no longer related to the sacred, so a cloud of forgetfulness developed between the inner and outer worlds, between the world of matter and the inner world of light. More recently this cloud has become denser and darker, attracting to it darker energies (for example, those embodied in corporate global exploitation) that made the inner light that should be guiding humanity more inaccessible. Because our attitude of consciousness affects the inner world, the greed that dominates our present culture, the focus on "me" as opposed to "we," has fuelled this gathering darkness. Our collective exploitation of our environment—which has resulted in irreversible climate change, mass extinction of species, and pollution—is also an expression of this attitude.

Tragically our Western civilization with its values of materialism and its forgetfulness of creation's sacredness has in the last few decades become a "global civilization." Its values are now global, and so are the effects of these values. Outwardly there is increased pollution and ecological devastation as more and more people want the "stuff" of materialism. Inwardly there is an increased alienation from the sacred alive in the core of creation.

What we don't understand is that humanity—just as an individual—can lose an opportunity. Through patterns of greed and forgetfulness, through a collective desecration, we have covered over and darkened the light in the inner world. Over time, the light has faded. What had been given has been lost. The light of the inner world that gave meaning and spiritual sustenance to our souls and to the whole world has been going out.

And now it is extinguished.

Something that for millennia was central to the inner life has gone, lost through our greed and arrogance, our ego-centered power-dynamics and forgetfulness of the sacred. We are not entering just an external era of extinction, but an inner dark age. And what is more dangerous is that we do not appear to know it is happening.

THE LOSS OF LIGHT

As the light of this previous era fades, so fades with it the spark that was given to help us transition to a new age. I described this spark in the introduction as a "light of oneness" that could have been used by humanity to affirm the divine unity of all life on this planet and build structures in the outer world that reflect this new consciousness.

There were indications of this possibility in our re-awakening to the interconnectedness of all of creation, a dawning of the consciousness of oneness, which is one of the central qualities of the next era. Together with this awareness we have already been given some of the tools and technologies of the new era, such as global connectivity through cell phones and the Internet. And yet, despite some implementation of "fair trade" and other sustainable practices, the development of "globalization" has just led to more exploitation and corporate greed, rather than values that are in service to the whole. Collectively we have created a greater divide between rich and poor, more ecological destruction, more collective forgetfulness of the sacred. We have placed short-term economic progress above real concern for the planet.

And so the light of this past era, rather than transforming into the light of the next era, has gone out, and the spark that was given as a catalyst for this evolution is no longer present.

What does it mean that the light of our era has gone out? The light carried the higher destiny and purpose of humanity and the world. It made possible transformation and evolution, on an individual and collective level.

Without this light there can be no real change, no shift in consciousness, no evolution, whatever our apparent intentions or aspirations. The light of the individual soul, our individual destiny, remains. But this individual light is no longer nourished or supported by the greater light of the era. Without the energy of this greater light the destiny of the individual soul, its possibility of evolution, is severely limited.

This is largely unknown in the West—how the individual soul needs and relates to the light of humanity. In our emphasis on individualism and separation, we have forgotten the interdependence of the individual and the whole. Individuals can not do it alone. We have forever depended upon a greater light that is present in the inner world. And now, as this greater light has gone out, so too has our own potential to evolve diminished.

And without this greater light, the individual soul is more susceptible to the darkness that is present in the world. This darkness is not just a passive force, not just the darkness of ignorance, but a darkness that can absorb our individual light. It has access through our small failings, our selfishness, and other negative characteristics. As the darkness grows, we easily regress.[9] It has become more and more easy to forget our divine nature and the sacred within life. It is more and more easy to get caught in the many illusions that the darkness fosters.

Although many individuals have embraced the new dimension of consciousness and its awareness of the sacred, this shift has not yet happened to our collective consciousness, to our corporate or political world. It is this collective shift that is needed if we are to restore and rebalance our inner and outer environment. But without the light of the era and the spark of oneness to help us co-create the future, collectively we cannot, at present, welcome in a new era; we cannot undergo a shift in consciousness towards oneness and the return to the sacred. We lack the light, the energy that is needed to make this happen. There is not the light and guidance to help us to make a meaningful transition.

I have seen this light going out over the last few years, but to finally see that the light had gone out was a shock. I don't think that it has happened before, at least not for many millennia.

And so we wait in the darkness of a dying world. We sense in our souls what we can see in the ecosystem, that something is over, that the world will not return to what it was. And the collective, still caught in its dream of materialism, feels an anxiety, even anger, as it knows that this dream has passed its sell-by date, that its promises of prosperity are empty.

THE WORK OF WITNESSING

As the planet faces an unprecedented future we need to be especially attentive and aware. We cannot afford to slip into forgetfulness, to be lulled asleep by the dreams of our dying culture. There is also an increased need for individuals to protect their light from the darkness. Without this light within the world to guide and protect us, it is more and more important to follow one's conscience, and engage in correct ethical behavior and attitude (in Sufism this is called *adab*), to make sure that one's individual light is not influenced or diminished by darkness.

For the same reason there is also an increased significance of the *sangha* or spiritual community that can provide a spiritual container to help to protect the light of the individual. We all need to work consciously to protect our own light through our spiritual practice and attitude, and to share this light with acts of loving kindness. In this simple way we can nourish and help each other during this very challenging time.

We must also remember that while the light of our era has been extinguished, this does not mean that the light of the Divine, the spiritual light within creation, has gone out. This divine light is present in every cell of creation. Without this light there could be no life, no existence. The whole of creation is like a single light from the Source that goes through a prism and becomes the many colors of existence.

Collectively, we have failed to honor creation's light, we have forgotten humanity's sacred role as guardians of the planet. But individually we cannot allow ourselves to turn from it. In our own ways it is important that we honor this light, this sacred role, despite a greater collective forgetting. It is especially important that we remember our responsibility towards the Earth, both outwardly and inwardly: that we live with ecological and spiritual awareness.

Lastly, there is a very important tradition of "witnessing." Many spiritual teachings talk about the importance of witnessing or watching without judgment or expectations. There is great power in consciously being aware of events without being caught in trying to fix or resolve them. At this time, it is important that we engage this capacity for witnessing. When the veils of perception are not clouded over through reaction and "problem solving," we see and allow things to be as they are. Only then can we take responsibility for what has taken place.

There is a pressing need for real awareness—to wake up to the reality of what we have done. We need to become aware of our present predicament in the inner and outer worlds. We cannot take responsibility for our present ecological imbalance, or know how to act, without recognizing the global devastation we have created. Nor can we take responsibility for what we have done to the inner worlds without becoming aware of the darkness and forgetfulness of the sacred that is covering us. Maybe this first step towards responsibility and accountability will help to redeem what we have desecrated. It is a step towards maturity and real awareness.

Together with the loss of the light, the sacred substance within creation is diminishing. What happens to the soul when it is no longer nourished by the sacred within creation?

HUNGRY GHOSTS *and the*
SACRED SUBSTANCE *in* CREATION
(2011)

Humanity evolves both collectively and individually. Without the light of an era there can be no collective evolution—there is no light to see the way, no energy to transform our collective consciousness. There is also another invisible tragedy at this present time that affects us more than we know. Our soul, our divine Self, comes into this world in order to have certain experiences. And there is in this world a certain substance that enables these life experiences to be sacred and thus deeply meaningful to the soul. In Sufism this substance is called the secret of the word *"Kun!"* ("Be!"). This sacred substance in creation enables the soul to have an experience *here* that is sacred, because if it is not sacred, it doesn't touch the soul—then our experiences in life do not help the soul to evolve. Then there is no deep meaning to our life.

The sacred substance in creation enables experiences in this world to be real, to be meaningful, to be part of the evolution of the soul. In traditional cultures this sacred substance in creation was nourished through the rituals of daily life—of baking bread, of weaving, of planting. This was, and is, central to all indigenous cultures, and it meant that everyday life was sacred. And because life was sacred then the soul could have a meaningful experience—and if the soul can have a meaningful experience it can grow and evolve.

For thousands of years the purpose of different civilizations was to look after this sacred substance, this spiritual dimension of creation—through rituals, ceremonies, prayer, and sacred music—so that the souls of people could be nurtured and they could have a meaningful life and their souls could evolve. For example, when the Pomo Indian people of Northern California wove baskets, the women would go out and pray over the grasses before they cut them. As they wove their baskets they would put the reeds or grasses through their mouths to moisten them, praying over them. The basket thus wove together the physical and the spiritual parts of life. All aspects of life were approached in this way, the warp and woof of physical and spiritual woven together into the single fabric of life that was never anything other than sacred. Indigenous peoples saw their life as a communion with earth and spirit that nourished them and at the same time nourished creation, the two being so interwoven it would not have been possible even to think of nourishing the one without nourishing the other.

But we have lost these sacred ways and forgotten their purpose. And we are present at a time when our collective culture has forgotten there is a sacred purpose to life, has forgotten that life has a sacred substance. We no longer look after this substance in creation; in fact we no longer even know that it needs to be looked after. A few cultures remain, like the Kogi of the Sierra Nevada de Santa Marta in Colombia, who work to maintain the balance of life, mediating between the material world and "Aluna," the cosmic consciousness that is the source of life and intelligence. These "Elder Brothers" gave a warning to us, the "Younger Brothers," that due to our treatment of the environment there is a great danger—because we don't know the damage we are doing.[1]

Now because of our forgetfulness of the sacred and our desecration of creation in the way we treat the environment, this sacred substance in creation is getting less and less accessible—it is almost becoming lost. I think this can also be seen in the way people find less and less meaning in the simple things in their life, and are more addicted to materialism and to the many surface distractions, because there is nothing deep that resonates. Now, what actually happens if the sacred substance in creation is lost or it becomes buried so deep the soul can't interact with it, is that we become what the Tibetan Buddhists call "Hungry Ghosts." Traditionally the "hungry ghost realm" is one of the six realms, whose creatures have empty bellies, small mouths, and scrawny, thin necks. They can never get enough satisfaction. They can never fill their bellies. They are always hungry, always empty. Our civilization's insatiable consumerism, which cannot fulfill our real nature, has made us live as "hungry ghosts," constantly desiring what cannot nourish us. And now, on the very deepest level, this is what our whole culture is moving towards—as our souls crave the sacred nourishment they can no longer access.

As this sacred substance diminishes, our souls can find less and less nourishment here. The worst-case scenario is the whole planet becomes a Hungry Ghost. Children will still be born, souls will still come into the world, but they will not be able to have a meaningful experience. And the simple joy that belongs to life as an expression of the Divine will have faded away. We will hardly remember that it was once central to life.

This is what happens when the sacred substance in creation is lost and any real purpose has gone. This is the cusp we are on at the moment—which is why it is not just an ecological crisis, it is a spiritual crisis. But the real danger of this spiritual crisis is that it is unreported, unrecognized, and we do not seem to be aware of what is really happening or its consequences. We are unaware of what we are losing.

Finally I was forced to recognize that, instead of the oneness I had longed for, humanity has chosen to live in an increasing divide.

DARKNESS *and* LIGHT
(SUMMER 2012)

A few years ago I came out of meditation with a sentence that disturbed and surprised me. The words I was given were simple: "Those who follow the light follow the light. Those who follow the darkness stay here." At that time the focus of my writing and teaching was on oneness: I was exploring how the next step in our evolution will be to awaken to the consciousness of oneness—the consciousness that we are all part of one interconnected spiritual organism. And here I was given a profound and almost paradoxical statement that said otherwise: that there would be a division between those who follow the light and those who follow the darkness. I was left with a strange sense of unease.

Over the years since I have meditated on this saying, trying to understand its message. Watching the outer world, listening within, and to the dreams and visions of others,

I have come to believe that we have arrived at such a time of division, of separation between darkness and light. Looking around at a world covered in materialism, wrapped in a profound forgetfulness of the sacred, there is little to indicate a world of light. Instead the forces of darkness—with their global exploitation and greed, with their unprecedented ecological destruction and desecration—appear to have taken over the world. Each time we have the potential to make a collective decision to help humanity and the environment, we have put exploitation before real care for our planet's well-being. In fact at the recent Rio Summit the term "sustainability" mutated into "sustained growth." Looking solely toward material progress, humanity has forgotten its sacred role as guardian of the planet.

And yet there are also signs of a global spiritual awakening. The principle of oneness, of being part of an interconnected living whole, is no longer a fringe idea. There is a longing within many people in different parts of the world to return to the sacred, to live a connection to the Divine within themselves and within life. Many are dissatisfied with the values of materialism and the soulless civilization it fosters. Individually and as groups people are aspiring to create a way of life based upon a relationship to what is sacred. Often it seems that those who aspire to live by this awakening light and those who are drawn deeper and deeper into the addictions of materialism are living on different planets.

Some people are waiting, hoping that new-age prophecies of the Mayan Calendar mean that the awaited global transformation will soon take place. Others see darker signs on the horizon, sensing that we are at a few minutes to midnight in a global ecological or economic collapse, an ecological "tipping point" whose effects we cannot predict. We all have our hopes and fears, some shared and some secret. But we cannot deny that primal changes are happening in the inner and outer worlds. Is this a crisis or an opportunity, or both?

My own sense and deep sadness is that we have reached or even passed the "tipping point" and that humanity has made its collective decision. It has decided to pursue its dreams of economic prosperity with disregard for the environment and its own soul. Collectively we have refused to take real responsibility for our actions, for the massive depletion of species, global warming, and the pollution that we continue to cause. Humanity has decided to remain in the darkness, forgetful of the sacred within all of life.

We cannot return to the simplicity of an indigenous lifestyle, or to the ways of the hunter-gatherer. But we could have made a transition towards a real sustainability that cares for all of life and its sacred interconnected nature. Instead we have made a collective decision for "sustained growth," regardless of its consequences. We have placed our own material welfare before the well-being of the whole. Our worship of the false gods of materialism has now become a globally destructive phenomenon.

What this means to the soul of humanity we cannot begin to understand. In its entire history humanity as a whole has never made such a collective decision—almost all previous cultures lived a deeply-rooted connection to the sacred, in whatever form that took; such a decision would have been unimaginable. And what is especially frightening now is our seeming lack of awareness of this choice or its possible consequences.

And now our Western consumer culture, which has no relationship to the sacred dimension of creation, has become a global monster. In our arrogance and ignorance we have decided to forget our sacred role as guardians of the planet. We have chosen not to listen to the cry of life itself, even as we poison its waters and bring its species to extinction. We do not even realize that there is also a spiritual dimension to our unprecedented ecocide that we are so heedlessly committing. Our desecration has created a vast inner wasteland. And the sacred substance within creation that gives deep meaning to our existence is hardly present anymore: humanity's soul is starving and we do not even notice. Caught in the darkness we have created, does humanity have any choice but to remain in this soulless world?

And yet this darkness is not all that is present. There is a light that is calling to us. Those who are responding look towards a different way of being—a way in which they are not seduced by materialism and its accompanying forgetfulness of the sacred. In many different ways they aspire to live this light of the Divine. But this light can no longer penetrate the darkness of our collective consciousness. It is more and more difficult for the light to exist in a world so devoid of the sacred—there is nothing to sustain it. The pathways of light have begun to separate from the world of darkness. The words that I was told those years ago begin to speak a truth I only now am starting to understand.

How can we access the light? The world of light is always present, only hidden by the desires of the ego and our instinctual drives. Our base instinctual drives, such as lust or aggression, are very tangible; some of our self-centered actions and attitudes are also very visible. Other seductions of the ego, though, are more subtle. For example, in our contemporary culture the ego has created an image of spiritual life that subtly serves the ego's own purposes. The spirituality that is now sold in our marketplace as a vehicle for self-improvement and self-empowerment, rather than service, has created a web of deception. Its promises, focusing on our ego's needs and desires, hide us from the simplicity of our real Self. The true light of the Divine cannot be marketed or sold, but like sunlight is free and belongs to all of us. However, it carries the price of consciousness—awareness undistorted by the ego. With this awareness comes the real responsibility that is always in service to what is highest.

If we are to be present in the light we need to turn away from the ego and its patterns of self-interest. Then we will find a very different world opening around us. The higher knowledge that belongs to the world of light will gradually become accessible. In this light we will see more clearly the oneness of our multidimensional world, the "unity of being" that the mystic has long known to be the real nature of existence. And we will learn how to live according to the ways of oneness, which are very different to the struggles and demands of duality—the apparent world of separation we have inhabited for centuries. The sacred will once again be present though in a new way—not as something to be sought or longed for, but something naturally within and around us—the Divine once again communing directly with us. And this relationship with the sacred will of necessity include creation: an awareness of the light of the Divine throughout the whole web of life. Once again we will be able to read nature's book of revelation and learn how to work with the real magic of creation. This is the future that is being offered to us at this moment in time.

At present these two realities appear to co-exist. People living in the same towns and the same streets are already in very divergent worlds. There are those who deny the dangers of climate change and think that we can safely continue our economic expansion. Believing in the myth of progress, they look to technology to save us—blind to any real consequences of our unsustainable way of life. The effects of the ecocide we are perpetuating and the accompanying desecration on the inner world of the soul do not even enter their consciousness. And there are many others who know a very different reality in which inner and outer sustainability, and a reawakening of the sacred, are living principles. How these communities will continue together in the future is uncertain.

Humanity has made its collective choice. Its leaders have endorsed short-term economic growth over any real care for the world. Choosing to remain in the darkness of unsustainable materialism, humanity has constellated a widening divide. The darkness is very present around us, and continues to entice us with its ego-centered desires and soulless distractions. The light offers us a new way of being, a new awareness of our divine nature. And yet, as with all transitions, we have to sacrifice something—our attachment to materialism, our focus on our self to the exclusion of the whole. We do not know how this collective loss of soul will affect humanity, nor how the Earth itself will respond to this split. But we need to know that something fundamental has changed, and that rather than entering an age of unity, there is a deepening divide between the light and the darkness.

A STORY *of a*
LOST OPPORTUNITY
(FALL 2012)

Telling this story brings a deep sadness to the surface. Maybe this opportunity of a global shift in consciousness, a collective awakening to oneness, was only a remote possibility. Maybe I was naïve in thinking that humanity could step out of the grip of all of the power structures, the multinational corporations, and other forces that are draining the lifeblood of our planet. Our patterns of collective greed and exploitation are too well established to succumb to the simplicity of a oneness that embraces all of life, that cares for all of creation. The powers of darkness are too dominant. And so the ecocide continues, and the deep desecration of the sacred in the inner and outer worlds accelerates.

Was this always our destiny? Could we have avoided this effect of our forgetfulness, our refusal to accept our role as guardians of the planet? Many of us felt the seeds of an awakening, and how our consciousness shifted towards the oneness that is central to our existence. We felt this energy within our self and within the world. Many individuals and groups still hold this light of a spiritual awakening. There is a sense of a "new story" of humanity being about oneness and interconnectedness—a lived relationship to the whole. And yet what confronts us is the reality of a ruthless destruction of the ecosystem, and collective forces that continue to refuse to take any real responsibility. And some of us have noticed a deepening split between the collective, which remains caught in this dream of material prosperity, and those who seek to live from a different center that recognizes life's oneness and respects the sacred within all of creation.

And this split, this unfolding drama of darkness and light, confronts us with the primal paradox: If everything is one, if the "unity of being" is the very nature of the fabric of life, why do we appear caught in this duality? Is it real, or just created by our conditioning? It is said that the first experience of consciousness in primitive man was the duality of light and dark, as day gave way to night. In the Judeo-Christian myth, consciousness came from eating the fruit of the tree of the knowledge of good and evil, which resulted in our expulsion from the essential oneness of the Garden of Eden. Consciousness brings duality, and from the awareness of duality comes choice and free will. In the undifferentiated oneness of the instinctual world, imaged as the serpent eating its tail (*ouroboros*), there is neither consciousness nor choice. The gift of consciousness brings duality and choice.

As I have mentioned, human evolution is essentially the evolution of consciousness, and the step humanity was being offered was to make the transition from the ego's world of separation into the awareness of a greater unity. Rather than the unconscious oneness of an instinctual belonging to the natural world, this would be a conscious awareness of oneness, in which we would recognize how each unique part belongs within a greater whole.[1] We would also become aware of the patterns of interrelationship that belong to life's evolving oneness. But to make this step we would have had to freely leave behind the focus on our own individual self. Once again a step in consciousness required choice, a choice that confronted us with the primal duality of light and darkness.

On the human stage an awakening to oneness includes the drama of duality and the freedom of choice. A primal change has happened within the world, and something foundational within life has shifted. But in order for this change to fulfill its real potential it needed the conscious participation of humanity. We needed to break free from the grip of the "I" and turn our attention to the whole. We were needed to work with life's emerging energy: an energy that belongs to life's unity. Instead we chose a different course.

Humanity as a whole remained too addicted to its gods of materialism, too caught in its images of self-interest, to respond. We stayed behind the wall we had long ago created, that isolates us from the world of light. Nor did we heed the warnings nature has sent us—we have not read the signs. We have now passed the "tipping point" in the inner and outer worlds, and do not know what will happen—we have no models for the future. All that can be seen clearly is the darkness of our denial.

Everyday life continues and in many ways it appears that nothing has changed. We still read of wars and the acts of terrorism. We appear always on the brink of economic collapse. And yet in the West our supermarket shelves remain stocked with produce from all over the world. Even amidst the dying of a dream we still seem surrounded by prosperity. Many feel a deep sadness within their soul and the soul of the world, but do not understand its real nature, do not know what we have lost.

The ways of light seem to have separated from the paths of darkness. Rather than any real global unity, there is this deep split between those who are awake to our primal need to take care of the world and those who continue to embrace its exploitation. The only real global unity is that we all live with the effects of this exploitation. Our cars and power stations continue to emit carbon, the global temperature continues to rise, and the soul of the world continues to cry. This is the reality of our present predicament. And we remain stranded in our apparent ignorance, not wanting to know what is really happening.

Maybe there is a deeper meaning to this darkness. As I mentioned in one of the earlier chapters, it too has a story to tell. But from a simple human perspective there is a need to honor the sadness of what we have lost, of the possibility for real transformation—how we could have worked together with the Earth and its emerging energies. Many years ago I was blessed, or cursed, to see what this could mean, the future that is waiting for us. I saw the possibility of a re-awakening of the sacred and life's magical nature, of a sense of the return of the Divine to everyday life and the deep pervasive joy this would bring. I also saw the technologies of the future, sources of power that did not pollute, knowledge that brought together the inner and outer worlds, the way of the shaman and the skills of the scientist.

This future is still waiting, but we now have to pay a steeper price for our actions, for our self-centered greed, for what we have done to the world. There is no way we can avoid the effects of our actions, of the pollution and desecration we have caused. We will have to learn to live in a world that is darker, where our soul is not nourished, where an inner beauty is not present. How long this will continue I do not know. Some dark ages have lasted for centuries. Maybe we will have to wait until the oil has run out, until the coal has been burned, until the sea levels have risen, or until our souls are so starving that something deep within humanity or the world rebels. The ancient energies of the Earth are still alive and we do not begin to understand how they are responding to both the energy of change and our collective resistance. But rather than attempting any prophecy I would continue to be aware of what each moment is telling us, watching the signs in the inner and outer worlds just as a sailor would read the winds and tides.

And from within this darkening there arises a cry that we hold the light that is left, the light that is within our self and within the spiritual body of the world. So much has been lost, so much has been desecrated by our endless desires, but those of us who are aware of the sacred need to hold what is left, hold it in our hearts and real awareness. The light of the sacred needs our care and protection. Maybe at some time it will give birth to the child with stars in its eyes, to the future whose seeds are still all around us. Without our relationship to this light nothing can be born, and the darkness will devour any real hope. Those of us who are aware of what we were given, of the oneness that was awakening, are needed to hold true to life's deeper purpose, the unfolding of the soul of the world. We need to stay attuned to the heart of the world and life's essential message of love, however the drama in the outer world unfolds.

But at this moment in time I sense a need to recognize, if only for an instant, the opportunity we have lost. We cannot afford to remain ignorant of what has happened, and we have to accept our responsibility. Maybe then we will begin to learn the lesson that life is teaching us. Maybe then we can return to our real work as guardians of the planet and our shared evolution.

EPILOGUE:
A WINTER SOLSTICE
(WINTER 2012)

The years pass, the seasons of the soul and the world soul come and go. And we are a part of this movement, these cycles that come from the very center of the cosmos. The final months of 2012 brought to North America a devastating hurricane—an omen of climate change—and another mass shooting. This time it was elementary schoolchildren and their teachers who were caught in the crosshairs of an increasingly dysfunctional culture, as if life itself were screaming at what is really happening. From our continuous news cycles we saw the images of floods and destroyed buildings, the pictures of the children. We may have registered the shock but could we really grasp the story that life was telling us?

And now, as the midwinter solstice passes, we have come to another end and beginning. From the ancient teachings of the Mayans[1] we have learned that a great cycle of time, the Long Count calendar, is completing and so once again a new beginning is possible. This ancient prophecy of the end of time has brought with it stories of an apocalypse, but in the images of our imagination we cannot recognize that we are the apocalypse, that our civilization is the monster destroying the world, tearing life up at the roots, our materialistic values and behavior destroying what is most precious in the inner and outer worlds. Instead all we can see are our fears, that our way of life might be threatened, that our dreams of economic prosperity might not be sustainable.

There is an end and a beginning; there is an ancient story being told. Life is dying and life is being reborn. And this prophecy about the cycles of time, associated with the astrological concept of a "galactic alignment," has a meaning. As I awoke early in the morning of the winter solstice, with a Pacific storm hammering against the windows, I felt within my heart and at the core of my soul the sense of a deep inner realignment, as if life was being reconnected with its deepest source. In this moment of cosmic time something essential to life was being rebalanced, and an ancient song could be heard again. This is the song of life, of creation and its primal purpose—the ancient song of belonging that links our soul and the world soul with the patterns of the cosmos. These are the deeper mysteries upon which the cycles turn.

I am by nature skeptical of any prophecies, and was inclined to dismiss my experience. But then a bald eagle came and sat on the top of a tree outside my window until I remembered the tradition of the land and recognized it as a messenger from the spirit world. It told me that this moment in cosmic time had a deep meaning for the Earth and all of its inhabitants, for the inner world of spirit and the outer physical world. When I had heard its message, it gracefully spread its wings and flew away towards the bay. These are the signs that come from where the worlds meet, where the mysteries of the inner world are spun into the outer. Such signs point to what is real.

How will this moment affect us? How will this re-alignment change our world? It is almost impossible to see or hear what is really happening as the surface clamor of the outer world consumes most of our attention. The deep listening that belongs to the shaman and the seer is not part of our culture. But I do feel that something fundamental has happened, something that belongs to the cosmic cycles of creation. Something essential has been given back to humanity and the whole world, and yet given in a new way.

The fact that this realignment came from the center of the galaxy gives a cosmic dimension to this song within creation. I have known for many years that the next step in our evolution would include a quality of cosmic consciousness, as humanity becomes aware of the deeper significance of its place not just within our solar system but within the greater cosmos. How this will unfold we do not know, except that there will be an inner and outer relationship between our planet and the cosmos. A time of isolation or separation within our own individual world will have come to an end. Rather than projecting this transition into science fiction images of aliens or galactic travel, I would begin with the microcosmic understanding of what it means to us individually to step outside of the isolated sense we have of our own separate self, and instead embrace and live from a consciousness of oneness in which we *know* how we are part of an interconnected whole.

This shift in consciousness is dawning around us, even if we have missed the opportunity to bring it into our collective consciousness. Stepping into this larger dimension of consciousness, which at the same time recognizes and values the unique significance of each part within the whole, brings us into a very different reality. We are present, interconnected with the whole at each moment. The cosmic dimension of this shift in consciousness has begun with an inner alignment with the center of the galaxy, as if our planet has once again found its place within the cosmos. A certain note has been struck, a certain resonance heard. Our Earth is brought back into a harmonic pattern within the galaxy. From within this harmonic pattern our global consciousness can evolve in a new way.

We know in our own individual story how our life changes when we are brought back into a sense of harmony within our self and our environment. The music in our life changes as we feel that we are in the right place, in the center of our life. Within this harmony healing can be given and the possibilities of life expand as we find our self more "in tune" with life. New meaning can come from our soul. The purpose of many sacred ceremonies and sacred buildings had this very intention—to bring us back into harmony with the sacred source of life and with our soul. The simple experience of watching the summer solstice sun rise between the standing stones at Stonehenge, or illuminate a stained-glass window in a cathedral, reminds us of our place within the sacred dimension of the solar system and helps to harmonize us with its greater energies. Being aligned with the energy at the center of the galaxy has a similar, if not more potent, effect.

And now, after this moment of the winter's solstice, life's real song is present again, carrying this cosmic note. Our soul and the soul of the world will feel it even if we do not notice. Our consciousness has been so censored that we can only see the outer play of events, and even then we rarely recognize what they are telling. We have lost the wisdom of how to interpret the dream of life, and we don't know how to be guided by its signs. We are hardly able to see anything except our own reflection in the mirror of our own illusions—the censored story our media tells us again and again, even though we know its promises are empty. Even though, when we hear of the senseless shooting of young children, we know that something is terribly wrong.

But in this moment of darkness, in this winter solstice, when it seems we have missed every opportunity, life is recreating itself anew. We are a part of life, part of this rec-reation, this realignment, even if our attention is completely distracted, even if our way of life is an agent of terrible destruction and desecration, exterminating species as it pollutes the inner and outer worlds. We are both spirit and matter, and along with all of creation we are being reborn. Distracted by the images on our televisions, computer screens, and now smart phones, we might not know this for generations. We are so busy we do not have time to witness what is really happening. There is so little light left it is hard to see, the noise of our daily life is so loud it is difficult to hear. But the cycles of life and the cosmos, the seasons of the soul and the world soul, continue. And the ancient promises are always kept, the promises between heaven and Earth, the promises that give us real hope and meaning, the promises that our souls can hear, even if our senses and our minds cannot.

NOTES

INTRODUCTION

1. By "inner worlds" I mean those worlds that are invisible to our physical sight but exist in other dimensions of reality: for example, the angelic world, the world of the *devas* or nature spirits, the archetypal world of symbols, the inner world of the soul and world soul. Shamans, mystics, and seers, among others, have traditionally had access to different inner worlds. Sadly, one of the greatest censorships of our Western culture has been to deny the existence of these inner worlds. This has been partly caused by scientific rationalism, but also by the Catholic Church, which persecuted those who had direct access to inner worlds, like the Gnostics.

2. In our human story a step in the evolution of consciousness often involves duality and choice. See p. 145.

3. This series of books includes: *Signs of God* (2001), *Working with Oneness* (2002), *Light of Oneness* (2004), *Spiritual Power* (2005), *Awakening the World* (2006), and *Alchemy of Light* (2007).

4. For further information on global consciousness and the role of the Internet, see: www.workingwithoneness.org/internet.

CH 1: A NEW LIGHT

1. "Intimations of Immortality from Recollections of Early Childhood," 1. 76. *Wordsworth Poetical Works*, ed. Thomas Hitchinson, (London: Oxford University Press, 1969).

CH 2: THE AWAKENING *of the* WORLD

1. For example, the Masters of Wisdom (known in the Sufi tradition as the *awiliyā*, or friends of God) traditionally look after the spiritual well-being of the world.

CH 4: A PRAYER *for the* WORLD

1. The deeper meaning of time is captured most beautifully in the passage from Ecclesiastes (3:1–8) that begins, "For everything there is a season, and a time for every matter under heaven."
2. This archetypal energy is often imaged as Chronos or Saturn, the Lord of Time.
3. The Findhorn Foundation in Scotland is an example of a community that worked in close cooperation with nature *devas*.
4. The significance of the naming of creation is present in the Qur'an, "And He taught Adam the names of all things" (*Sūra* 2:31).
5. I have written a number of books, including *Light of Oneness* and *Alchemy of Light*, that contain some of the teachings about spiritual light that belong to the future.
6. See Vaughan-Lee, *Awakening the World*, pp. 115–124.
7. *Macbeth*, Act 5, Scene 5, 24–28.
8. Qur'an 24:40.
9. This is similar to Krishna's prophecy in the *Bhagavad-Gita*:

    Whenever dharma declines and the purpose of life is forgotten,
    I manifest myself on earth.

CH 6: THE DARKNESS BEFORE *the* DAWN

1. *The Second Coming.*
2. The archetypes are like the riverbeds through which the waters of life flow. They are the primal organizing patterns underlying creation. These universal patterns or motifs, which come from the collective unconscious, are also the basic content of religions, mythologies, legends, and fairytales. They emerge in individuals through dreams and visions. Irrepresentable in themselves, their effects appear in consciousness as the archetypal images and ideas.

CH 7: WHAT HAPPENS WHEN *the* ICE MELTS

1. Interestingly, Isaac Newton himself was an alchemist, leading John Maynard Keynes, who acquired many of Newton's writings on alchemy, to state, "Newton was not the first of the age of reason. He was the last of the magicians."

2. In complete contrast are the teachings of Zen, as expressed by Zen master and hermit Stonehouse at the beginning of a retreat: "From dawn to dusk, whenever you lift your feet or put your feet down, you may not step on permanent ground." (*The Zen Works of Stonehouse*, trans. Red Pine, p. 171.)

3. For a fuller understanding of the symbolic consciousness of the medieval period, see C.S. Lewis, *The Discarded Image: An Introduction to Medieval and Renaissance Literature.*

4. Symbolic consciousness has been described by Carl Jung as "mythological thinking." It allows for the formation of symbols, and as such, for a symbolic relationship to life. It is older than analytic thinking and is fundamentally subjective, pre-verbal, and mythological.

5. In the same way, Western colonialists imposed their "enlightened" views upon "primitive" indigenous peoples.

6. Qur'an 41:53.

CH 8: THE WALL

1. Some children still have direct access to this world of light, before they become caught in the confines of adult consciousness.

> *There was a time when meadow, grove and stream,*
> *The earth, and every common sight,*
> *To me did seem*
> > *Apparelled in celestial light,*
>
> …
>
> *Heaven lies about us in our infancy!*
> *Shades of the prison-house begin to close*
> *Upon the growing Boy.*

(William Wordsworth, "Intimations of Immortality," ll. 1–5, 66–8.)

2. Indigenous cultures that did not go against the natural law did not seek to have power over nature, did not experience this split between spirit and matter.

3. Traditionally humanity was supposed to act as "vice-regent" of God, His representatives here on Earth, rather than usurping all power and authority for themselves. The ruler as "priest-king" embodies this spiritual role as mediator between heaven and Earth before the "Fall."

4. Christian suppression of paganism was begun by Constantine I, but it was Theodosius who issued a law that prohibited any type of pagan worship and authorized the destruction of many temples throughout the Roman Empire.

5. In the Bible (Genesis 2:20) Adam was authorized by God to give the creatures of creation their names: "And Adam gave names to all of the cattle, and to the fowls of the air, and to every beast of the field." In Hebrew *adam* means "human," and in Sufism *Adam* is the essential human being.

6. According to Ibn 'Arabī this knowledge of the names was transmitted through the succession of perfect human beings: "The perfect human beings never ceased receiving the names from one another until the names finally reached Muhammad." (Chittick, *The Self-Disclosure of God*, p. 154.)

7. Lyrics from the song "Anthem."

8 The Earth is a living, spiritual being, with its own soul, understood by the ancients as the *anima mundi*.

CH 9: WITNESSING *the* END *of an* ERA

1. H. Richard Lane, program director in the National Science Foundation's Division of Earth Sciences.

2. See: www.ecobuddhism.org/wisdom/interviews/jmacy. Since Joanna Macy gave this interview, it has become more acknowledged that we may be reaching the "tipping point" of global climate change.

3. Martin Prechtel, "Saving the Indigenous Soul," *The Sun*, April 2001.

4. For example, *adab*, or spiritual courtesy, is central to Sufism. *Adab* is an outer expression of an inner spiritual attitude.

5. Carl Jung describes this tragedy: "Man himself has ceased to be the microcosm and his anima is no longer the consubstantial *scintilla* or spark of the *Anima Mundi*, the World Soul." *(Collected Works*, vol. 11, ¶ 759.)

6. In Sufism this is the mystery of "light upon light," whereby the light of our aspiration attracts a higher, divine light that helps us on our journey.

7. For further information on the evolution of human consciousness, see the Joseph Campbell series *The Power of Myth*, which explores how the development of humanity is embodied in its mythology. In particular, in this era when the image of the Earth from the lunar landings was published, it led to the universal realization that humanity must identify with the entire planet, and the emergence of a new mythology based on global aspects of life.

8. The knowledge of sacred geometry was central to the cathedral builders. At Chartres there was a mystery school that combined the knowledge of music, harmonics, and sacred geometry, as well as other esoteric disciplines. Sacred geometry can align or "harmonize" an individual with his or her inner spiritual self, and also align the inner and outer on a larger, macrocosmic scale.

9. This has been most evident when people are caught in a collective darkness—as for example Nazi Germany, or more recently Bosnia or Rwanda—and ordinary people regress into terrible cruelty.

CH 10: HUNGRY GHOSTS *and the* SACRED SUBSTANCE *in* CREATION

1. "Up to now we have ignored the Younger Brother. We have not deigned even to give him a slap. But now we can no longer look after the world alone. The Younger Brother is doing too much damage. He must see, and understand, and assume responsibility. Now we will have to work together. Otherwise, the world will die." For further information on the Kogi, visit: www.taironatrust.org.

CH 12: A STORY *of a* LOST OPPORTUNITY

1. The evolution of consciousness is thus a journey from oneness to oneness: from an unconscious, instinctual oneness, through awakening of individual consciousness and its accompanying experience of duality and separation, to a conscious awareness of the oneness to which we have always belonged. To quote T.S. Eliot, from "Little Gidding":

> *And the end of all our exploring*
> *Will be to arrive where we started*
> *And know the place for the first time.*

EPILOGUE: A WINTER SOLSTICE

1. It is important to recognize that the Mayan culture had a deep understanding of time's astrological cycles and their spiritual significance. Our culture has lost any understanding of the spiritual dimension of time and its cycles.

ACKNOWLEDGMENTS

For permission to use copyrighted material, the author gratefully wishes to acknowledge:

Lines from *The Zen Works of Stonehouse*, Copyright © 1999 by Red Pine, reprinted by permission of Counterpoint; Extract taken from 'Little Gidding' in *Four Quartets*, (c) Estate of T.S. Eliot and reprinted by permission of Faber and Faber Ltd.; and Excerpt from "Little Gidding" from *Four Quartets* by T.S. Eliot, Copyright © 1942 by Houghton Mifflin Harcourt Publishing Company, Copyright © renewed 1970 by T.S. Eliot, reprinted by permission of Houghton Mifflin Harcourt Publishing Company, all rights reserved.

ABOUT THE AUTHOR

LLEWELLYN VAUGHAN-LEE, Ph.D., is a Sufi teacher in the Naqshbandiyya-Mujaddidiyya Sufi Order. Born in London in 1953, he has followed the Naqshbandi Sufi path since he was nineteen. In 1991 he became the successor of Irina Tweedie, author of *Daughter of Fire: A Diary of a Spiritual Training with a Sufi Master*. He then moved to Northern California and founded The Golden Sufi Center (www.goldensufi.org). Author of several books, he has specialized in the area of dreamwork, integrating the ancient Sufi approach to dreams with the insights of Jungian Psychology. Since 2000 the focus of his writing and teaching has been on spiritual responsibility in our present time of transition, and an awakening global consciousness of oneness (www.workingwithoneness.org). More recently he has written about the feminine, the *anima mundi* (world soul), and spiritual ecology (www.spiritualecology.org).

ABOUT THE PUBLISHER

THE GOLDEN SUFI CENTER publishes books, video, and audio on Sufism and mysticism. A California religious nonprofit 501 (c) (3) corporation, it is dedicated to making the teachings of the Naqshbandi Sufi path available to all seekers.

For further information about activities
and publications, please contact:

THE GOLDEN SUFI CENTER
P.O. Box 456 · Point Reyes Station · CA · 94956-0456
tel: 415-663-0100 · fax: 415-663-0103
www.goldensufi.org

THE GOLDEN SUFI CENTER PUBLICATIONS
www.goldensufi.org/books.html

OTHER BOOKS *by* LLEWELLYN VAUGHAN-LEE

SPIRITUAL ECOLOGY:
The Cry of the Earth

PRAYER OF THE HEART
IN CHRISTIAN AND SUFI MYSTICISM

FRAGMENTS OF A LOVE STORY:
Reflections on the Life of a Mystic

THE RETURN OF THE FEMININE
AND THE WORLD SOUL

ALCHEMY OF LIGHT:
Working with the Primal Energies of Life

AWAKENING THE WORLD:
A Global Dimension to Spiritual Practice

SPIRITUAL POWER:
How It Works

MOSHKEL GOSHA:
A Story of Transformation

LIGHT OF ONENESS

WORKING WITH ONENESS

THE SIGNS OF GOD

Love is a Fire:
The Sufi's Mystical Journey Home

The Circle of Love

Catching the Thread:
Sufism, Dreamwork, and Jungian Psychology

The Face Before I Was Born:
A Spiritual Autobiography

The Paradoxes of Love

Sufism:
The Transformation of the Heart

In the Company of Friends:
Dreamwork within a Sufi Group

The Bond with the Beloved:
The Mystical Relationship of the Lover and the Beloved

⁓

EDITED *by* LLEWELLYN VAUGHAN-LEE
with biographical information by SARA SVIRI

Travelling the Path of Love:
Sayings of Sufi Masters

⁓